Engineers for Change

Engineering Studies series
edited by Gary Downey

Matthew Wisnioski, *Engineers for Change: Competing Visions of Technology in 1960s America*

Engineers for Change
Competing Visions of Technology in 1960s America

Matthew Wisnioski

The MIT Press
Cambridge, Massachusetts
London, England

MIT Press books may be purchased at special quantity discounts for business or sales promotional use. For information, please email special_sales@mitpress.mit.edu or write to Special Sales Department, The MIT Press, 55 Hayward Street, Cambridge, MA 02142.

This book was set in Sabon by Toppan Best-set Premedia Limited, Hong Kong. Printed and bound in the United States of America.

Library of Congress Cataloging-in-Publication Data

Wisnioski, Matthew H., 1978–
Engineers for change : competing visions of technology in 1960s America / Matthew Wisnioski.
    p.   cm. – (Engineering studies series)
Includes bibliographical references and index.
ISBN 978-0-262-01826-5 (hardcover : alk. paper)
1. Technology–Social aspects–United States. 2. Technology–United States–Forecasting.
I. Title.
T14.5.W5664   2012
303.48'3097309046–dc23
2012009890

10   9   8   7   6   5   4   3   2   1

"We live in a world of change," to quote Adam as he led Eve out of the Garden of Eden.

—Augustus Braun Kinzel, "Engineering and Our Way of Life"

Contents

Series Foreword

We live in highly engineered worlds. Engineers play crucial roles in the normative direction of localized knowledge and social orders. The Engineering Studies series highlights the growing need to understand the situated commitments and practices of engineers and engineering. It asks: What is engineering for? What are engineers for?

Drawing from a diverse arena of research, teaching, and outreach, the Engineering Studies series raises awareness of how engineers imagine themselves in service to humanity, and how their service ideals impact the defining and solving of problems with multiple ends and variable consequences. Books in this series examine relationships along technical and nontechnical dimensions, and how these relationships change over time and from place to place. The researchers often are critical participants in the practices they study.

The Engineering Studies series publishes research in historical, social, cultural, political, philosophical, rhetorical, and organizational studies of engineers and engineering, with particular attention to normative directionality in engineering epistemologies, practices, identities, and outcomes. Areas of concern include engineering formation, engineering work, engineering design, equity in engineering (gender, racial, ethnic, class, geopolitical), and engineering service to society.

The Engineering Studies series thus pursues three related missions: (1) advance the understanding of engineers, engineering, and outcomes of engineering work; (2) help build and serve communities of researchers and learners in engineering studies; and (3) link scholarly work in engineering studies to broader discussions and debates about engineering education, research, practice, policy, and representation.

Matthew Wisnioski's *Engineers for Change: Competing Visions of Technology in 1960s America* is a wonderful book for launching the series. Calling critical attention to engineers' claimed jurisdiction over technology, it highlights the struggles of engineering intellectuals during the 1960s to reimagine that jurisdiction when it was no longer possible to view technology as inherently progressive. How those struggles helped develop and promote the broader vision of what Wisnioski calls the "ideology of technological change" serves as a cautionary tale for anyone who seeks to challenge and intervene in dominant images of engineering practice.

Welcome to the Engineering Studies series!

Gary Downey

Abbreviations

AEG	Allgemeine-Elektricitäts-Gessellschaft
AIAA	American Institute of Aeronautics and Astronautics
AIChE	American Institute of Chemical Engineers
AID	Agency for International Development
ASCE	American Society of Civil Engineers
ASEE	American Society for Engineering Education
ASME	American Society of Mechanical Engineers
AT	Appropriate Technology
AT&T	American Telephone and Telegraph Company
BART	Bay Area Rapid Transit System
BBN	Bolt, Beranek and Newman
BNL	Brookhaven National Laboratory
Caltech	California Institute of Technology
CAVS	Center for Advanced Visual Studies
CESO	Council of Engineers and Scientists Organizations
CIA	Central Intelligence Agency
CITP	Civilian Industrial Technology Program
CPA	Center for Policy Alternatives
CSIT	Committee on Social Implications of Technology
CSRE	Committee for Social Responsibility in Engineering
DOD	Department of Defense
E.A.T.	Experiments in Art and Technology
ECPD	Engineers' Council for Professional Development
EDP	UCLA Educational Development Program
EJC	Engineers Joint Council
EPA	Environmental Protection Agency
ESJP	Engineering, Social Justice, and Peace

EWB-USA	Engineers Without Borders
FASST	Fly America's Supersonic Transport (later, Federation of Americans Supporting Science and Technology)
GE	General Electric Company
GM	General Motors Corporation
HMC	Harvey Mudd College
IBM	International Business Machines Corporation
ICBM	Intercontinental Ballistic Missiles
IDA	Institute for Defense Analyses
IEEE	Institute of Electrical and Electronics Engineers
IIT	Illinois Institute of Technology
IST	International Science and Technology
LSD	Lysergic acid diethylamide
MASE	Mohawk Association of Scientists and Engineers
MIT	Massachusetts Institute of Technology
NAE	National Academy of Engineering
NAS	National Academy of Sciences
NASA	National Aeronautics and Space Administration
NPF	National Professions Foundation
NRC	National Research Council
NSF	National Science Foundation
NSPE	National Society of Professional Engineers
NYRB	New York Review of Books
OEO	Office of Economic Opportunity
ONR	Office of Naval Research
OTA	Office of Technology Assessment
R&D	Research and Development
RANN	Research Applied to National Needs
RPI	Rensselaer Polytechnic Institute
SACC	Science Action Coordinating Committee
SDS	Students for a Democratic Society
SftP	Science for the People
SHOT	Society for the History of Technology
SSRS	Society for Social Responsibility in Science
SST	Supersonic Transport
STS	Science, Technology, and Society

T&S	Technology and Society
TRW	Thompson Ramo Wooldridge Inc.
UCLA	University of California, Los Angeles
UCS	Union of Concerned Scientists
UN	United Nations
VITA	Volunteers for International Technical Assistance (later, Volunteers in Technical Assistance)

Acknowledgments

When the network of advocates for an intellectual project crosses a threshold of scale, we resort to stories of crystallization or eternal nature. The historian's task is reversal, to show that the "idea" always was an entanglement of human relations.

As series editor, colleague, mentor, and friend, Gary Downey has made the past five years the most exciting of my career. I see the results of our conversation on every page. Ellsworth Fuhrman and the Virginia Tech Department of Science and Technology in Society provided critical intellectual and financial support. Audra Wolfe disciplined my voice, and Monique Dufour prepared it for the main event. Marguerite Avery, Katie Persons, and the staff at MIT Press were patient and encouraging as the book grew in response to excellent anonymous referees, and Dana Andrus was a swift and insightful editor.

Engineering studies is a flexible concept, benefiting from the intersection of multiple communities. Material was tested at the Colorado School of Mines, Columbia University, Drexel University, Dublin Institute of Technology, École normale supérieure de Cachan, Harvey Mudd College, two International Network for Engineering Studies workshops, Lafayette College, Massachusetts Institute of Technology, Northwestern University, Princeton University, Radcliffe Institute for Advanced Study, Royal Institute of Technology, Society for the History of Technology's Woods Hole writing workshop, Stanford University, University of California Berkeley, University of Virginia, University of Wisconsin Madison, Virginia Tech, Washington University in St. Louis, and annual meetings of the American Historical Association, History of Science Society, and Society for the History of Technology. For organization and commentary I thank: Atsushi Akera, Ken Alder, Garland Allen, D. Graham Burnett, Benjamin

Cohen, Maja Fjæstad, Philippe Fontaine, Anders Houltz, Timothy Hyde, Alastair Iles, Caroline Jones, David Kaiser, Scott Knowles, Juan Lucena, Erika Milam, Carl Mitcham, Kelly Moore, Tania Munz, Mike Murphy, Dean Nieusma, Dick Olson, Dan Plafcan, Susan Silbey, Rebecca Slayton, Laura Stark, John Staudenmaier, James Tejani, Rosalind Williams, and Langdon Winner. Many other scholars have read and commented in part or in full, including: Ross Bassett, Brooke Blower, John K. Brown, Jamie Cohen Cole, Angela Creager, Chris Csikszentmihályi, Deborah Douglas, Arindam Dutta, Mayanthi Fernando, Matthew Gill, Michael Gordin, Andrew Jamison, Brent Jesiek, Ann Johnson, Aditya Johri, Ron Kline, Patrick McCray, Cyrus Mody, Joseph November, Guy Ortolano, Donna Riley, Eric Schatzberg, Sonja Schmid, Bruce Sinclair, Amy Slaton, Alistair Sponsel, Corinna Treitel, and Adelheid Voskhul. Dan Bouk and Molly Loberg deserve special recognition for reading multiple iterations. Tristan Cloyd and Whitney Schaefer gave research assistance, and Jongmin Lee, Sumitra Nair, and Nicholas Sakellariou their comments. Students in my Engineering Cultures classes helped polish the argument.

Without teams of archivists and assistants this book would not have been written. I thank Judith Goodstein and Shelley Erwin at the California Institute of Technology libraries; Carrie Marsh at the Claremont College Library; the staff of the Getty Research Institute, Harvard University Archives, Princeton University Archives, and University of Utah Archives; Wendy Chmielewski at the Swarthmore College Peace Collection; and Charlotte Brown at the UCLA Archives. Nora Murphy and the staff of the MIT Institute Archives and Special Collections were especially helpful over multiple visits. Donald Browne at UCLA gave access to the steel shed on top of the School of Engineering and Applied Science.

Scholars, engineers, and artists have shared documents, memories, and reproduction permission, including: J. Malvern Benjamin, Larry Bucciarelli, David Drew, David Fradin Samuel Florman, Joshua Lerner, Julie Martin, Don McAllister, Usman Mushtaq, Noel de Nevers, Gale E. Nevill Jr., Stephen Unger, Charles Weiner, Gershon Weltman, and Bess Williamson. The following organizations also graciously granted image permissions: the American Society of Mechanical Engineers, American Association for the Advancement of Science, Engineers Without Borders-USA, Engineers for Social Justice and Peace, General Electric Company, Institute of Electrical and Electronics Engineers, MIT Museum, Roy Lichtenstein Foundation, and Singularity University.

Taylor and Francis Ltd. allowed me to reprint material from the article "Inside the System: Engineers, Scientists, and the Boundaries of Social Protest in the Long 1960s," *History and Technology* 19, no. 4: 313–33 (copyright © 2003) in chapter 4. And portions of chapter 6 first appeared as "'Liberal Education Has Failed': Reading Like an Engineer in 1960s America," *Technology and Culture* 50, no. 4 (October): 753–82 (copyright © 2009 by The Johns Hopkins University Press, used by permission).

A two-year Andrew W. Mellon fellowship at Washington University's Modeling Interdisciplinary Inquiry Program under the mentorship of Steven Zwicker and Howard Brick proved vital for re-conceptualizing the book and expanding its evidence. At Princeton, the project emerged out of the productive tension of two would-be engineers, Michael S. Mahoney and Daniel T. Rodgers. I hope Mike would have liked it. At the Johns Hopkins University, Stewart W. Leslie, Pamela Long, David Munns, and Larry Principe set me on this path.

No one has done more to hold the project and its author together than Cindy Rosenbaum. If writing a book is a labor of love, this one was hers. No one made it more possible than my parents, for whom there are no adequate thanks. Isaak, please forgive your dad his trips to the coffee shop.

When I was twenty, I had the opportunity to work at the Army Research Laboratory in a partnership with the Johns Hopkins University. James B. Spicer, Chris Richardson, J. Derek Demaree, and James K. Hirvonen shared the challenges and pleasures of their world. It has taken me a decade to appreciate fully what they gave. This book is for them.

1

Introduction

My mind was lofty and wished for a new world and way of life . . .
—*Menocchio*[1]

The pamphleteers in Columbus Circle warned of technology run amok as thirty thousand dark-suited men streamed into the New York Coliseum for the world's largest gathering of engineers. "Technology was invented to serve man," their leaflets harangued, "and yet everywhere his needs are at a crisis." Smiling under the nose cone of a cartoon H-bomb, a hastily scribbled David Packard—the 1971 Institute of Electrical and Electronics Engineers (IEEE) International Convention's keynote speaker—demanded the passerby's attention. "We believe that the current misuse of our technology can be turned around," it was printed a thousand times over, "but not if we set up David Packard and others like him as our examples."[2] Denied a booth inside the Convention, the dissident engineers worked to direct "productive and creative people" to their counterconference a few blocks away at the Ethical Culture Society Hall.[3] Those who made the trek discussed how to turn "necrophilic" technologies into "humane" ones in a new organization called the Committee for Social Responsibility in Engineering (CSRE).[4] They learned the structural causes of the recession among technical workers from Columbia University professor Seymour Melman and Congressman Ed Koch, while Quaker engineers Victor Paschkis and J. Malvern Benjamin described how "individual freedom and responsibility" were compatible with professional practice. Before heading back to the Coliseum, visitors could grab an inaugural issue of *Spark* magazine, which asked for help building a national movement of responsible technologists.[5]

Figure 1.1
David Packard as depicted on CSRE leaflet
Source: CSRE, "Packard???" *Spark* 1, no. 1 (spring 1971): 46. Courtesy of
Stephen H. Unger.

At the convention banquet in the nearby New York Hilton, Packard,
the man the CSRE deemed "Mr. Military/Industrial Complex," stepped
to the podium, where three silent protesters from the group Scientists
and Engineers for Social and Political Action stood holding signs. The
demonstrators did not faze Packard; in fact he began by reading their
leaflets aloud.[6] He needed no lesson in responsibility. With only $538 in
initial capital he and William R. Hewlett had built a $300 million
company employing sixteen thousand employees that tangibly contributed to technological and human progress, a company from which he
took a leave of absence to serve as President Richard Nixon's deputy
secretary of defense.[7] Besides, compared to the terrorists who firebombed
his partner's Palo Alto home just two months earlier, what influence
could a handful of disgruntled professionals exert?[8] At a keynote session
the previous evening—"Redirecting Electro-Technology for a Better
World"—a panel of distinguished colleagues that included Edward E.
David Jr. of the Office of Science and Technology and J. Herbert Hol-

lomon of the Massachusetts Institute of Technology (MIT) already had warmed the crowd to the leadership opportunities engineers need to take to apply technology to social problems. Whether for national defense or domestic prosperity, Packard assured, research and development (R&D) programs would be improved and the waste and false promises of "brochure engineering" would give way to smart solutions.[9]

This book addresses deep-seated assumptions about technology and American life by revisiting a moment when engineers were at war with their ideals. Between 1964 and 1974 a rift about the purposes of engineering and the nature of technology opened within the profession, sparked by a combination of changes in the organization, content, and scale of engineering labor, and by a trenchant critique of technology from intellectuals, activists, and everyday people. The most significant outcome of this crisis, I will argue, was how, largely in reaction to the alternative futures of reformers, engineers adopted a powerful vision of autonomous technological change that continues to shape how their profession, and indeed the majority of Americans, encounter the human-built world.

To be an American engineer in the aftermath of World War II had been to look upon a seemingly limitless future. While the idea that material progress inherently brings social progress always has had challengers, in the United States, the period from 1945 to 1964 was one of near-utopian belief in technology's beneficence. Few denied that the nation was undergoing a scientific revolution, as popular imagery portrayed a futurist world of flying cars and plastic houses.[10] In democracy's name, engineers found government patronage on the frontiers of electronics, aeronautics, and nuclear power that swelled the profession's ranks.

And yet, in the decade that followed, technology took on ambiguous and ultimately sinister connotations in American thought and culture. "Technique has become autonomous," the Christian anarchist Jacques Ellul wrote in the 1964 English translation of his *Technological Society*, "it has fashioned an omnivorous world which obeys its own laws and which has renounced all tradition."[11] That same year Frankfurt school émigré Herbert Marcuse claimed in *One Dimensional Man* that technology curtailed, rather than accelerated, productive social change in modern society.[12] Not to be outdone, Lewis Mumford, one of America's foremost public intellectuals, warned: "With this new 'megatechnics' the dominant minority will create a uniform, all-enveloping, super-planetary structure, designed for automatic operation."[13] According to their commentary, as a result of technological imperatives—not just the accretion of material inventions but the systematic interlocking of artifacts, organizations, and patterns of efficient behavior—contemporary life was becoming more alienating, more destructive, more totalitarian, and less human.

By 1968, strands of critical thought provided commonality between countercultural, environmental, civil rights, and antiwar movements in what historian Theodore Roszak characterized as "a cultural constellation that radically diverges from values and assumptions that have been in the mainstream at least since the Scientific Revolution of the seventeenth century."[14] Never before had technological power appeared simultaneously so autonomous and so inextricable from political power, and not since the machine-breaking uprisings of the early nineteenth century had so many citizens perceived technology as a force to be resisted.

Engineers, it might be assumed, played no part in the social movements of the 1960s, much less called into question technology's progressive worth. In most theorists' conceptions, engineers were the embodiment of the military-industrial complex: conformist organization men in the system that stood to be torn down. But as the protests at the IEEE Convention attest, environmental degradation, the Vietnam War, and a host of socio-technical concerns led reformers to pressure their profession to honor its social responsibilities. And yet we see in Packard's speech that professional elites likewise confronted the proper responsibilities of the nation's technical labor force. To restore progressive worth to their profession, engineers recast a range of values—*responsibility, creativity,* and *change*—in competing visions of the future between which debates were fought, organizations patterned, and technical decisions made.

The seeds of this book took root when as a student I encountered two classic social studies of engineering at the same time. In 1971, the historian Edwin T. Layton Jr. published a book of shared interest to engineers on the streets and in the banquet hall. His *The Revolt of the Engineers: Social Responsibility and the American Engineering Profession* documented the origins of a professional vision he called "the ideology of engineering." With the rise of science-based industry, engineering became a mass occupation whose practitioners were subject to conflicting values of science and business. Desiring autonomy but subject to corporate bureaucracy, a small cadre of engineers in the 1910s and 1920s appropriated notions of social responsibility from the era's Progressive movement. These reformers portrayed themselves as technology's masters in service to the public, arguing that "the future society would be what the engineering profession willed it to be."[15] While they waged an unsuccessful attempt to wrest control of their professional organizations from business interests and failed to implement technocratic governance, remnants of their ideology proved lasting.

The *Revolt of the Engineers* was a pioneering text in engineering studies that generated scholarly debates about how to analyze engineers as political actors. By focusing on the profession's visionaries, Layton highlighted the role of ideology and identity politics in engineering practice.[16] But I became captivated by *Revolt of the Engineers* for the same reason as the president of the American Society of Mechanical Engineers (ASME), who, in his 1971 review of Layton's book, concluded that "many of the papers and speeches cited could have been presented in the last few years without loss of impact."[17]

While engaging with Layton's analysis, I came upon another of his many readers in the engineering profession. In 1976 Samuel C. Florman turned popular attention to the malaise plaguing America's technologists with his crossover literary hit, *The Existential Pleasures of Engineering*. "What is it like to be an engineer at the moment that the profession has achieved unprecedented successes," Florman asked, "and simultaneously is being accused of having brought our civilization to the brink of ruin?"[18] In a requiem to unabashed technological optimism, he chronicled the waning of engineering's "Golden Age," laid low by "anti-technologists" who had channeled concerns over nuclear weapons, corporate alienation, and environmental accidents into a dystopian worldview. According to Florman, a near decade-long "frenetic evangelical crusade" to redirect technology to idealistic social purposes had generated only illusory panaceas.[19] Americans consequently were left questioning technology's value and man's ability to control it. For engineers, the situation was grim. What authority could they claim over technology when confronted with such transparent failures?

To Florman and fellow engineers the anti-technologists challenged the essence of what it meant to be an engineer. In the breaking of the narrative of progress, engineers lost their already tenuous image as technology's masters. More than any particular socio–technical problem, these engineers fixated on the insurgency that was undermining their legitimacy and that of their employers. In professional member societies they decried an onslaught of "radicals," "pessimists," "hippies," "cynics," "anarchists," "environmentalists," "feminists," "militants," "modern 'witches'," "dropouts," and "revolutionaries." The "intellectual Luddites" with their "shrill voice of dissent" cast in "high-flown drivel," "immature emotional thinking," "hysterical outcries," "radical liberal clap trap," and "filthy speech" were "sapping our national vitality" and had cornered engineers into a situation "like Lavoisier to the guillotine."[20]

Almost every attack on antiwar protesters, hippies, and "so-called intellectuals," however, provoked a curious defense within engineering's own ranks. In 1972, while Florman was working out his argument for *Existential Pleasures*, he published a comparative review in the journal *Civil Engineering* of "two misleading and dangerous works" from "glib and simple-minded prophets"—Ellul's *Technological Society* and the biologist René Dubos's Pulitzer Prize winning *So Human an Animal*.[21] While one American Society of Civil Engineers (ASCE) member boasted that he was sending Florman's "penetrating and brilliant analysis" to *Reader's Digest*, another chided Florman's approach: "I would hesitate to stand up in a public hearing, trying to justify a major program of public works, and confront my adversaries, who may have read Ellul and Dubos and maybe even Lewis Mumford, by using language such as Mr. Florman has found appropriate."[22] Another, who implored his fellow ASCE members to seek out a veritable bibliography of critical texts, was less kind: "As between the Ellul and Dubos books on the one hand and Mr. Florman's own philosophy on the other, the latter appears to me much more deserving to be called 'misleading and dangerous.' . . . One can almost hear a chorus of physiocrats in the wings, chanting 'Laissez-faire, laissez-faire!'"[23]

Dissent among engineers often amounted to furtive acts of individual resistance—wearing an antiwar armband at work or penning an invective letter to the editor of a technical journal—but a small and vocal minority attempted to redefine engineering by rethinking the nature of technology through collective action. In both the ASME and the IEEE, engineers established "technology & society" committees and worked to bolster ethical codes. On college campuses faculty converted defense research to domestic needs. Corporate employees collaborated with artists to assert that human creativity was integral to the design of technical artifacts.

Discontent with the cold war order extended moreover beyond leftist academics, religious moralists, and iconoclasts. The construction of the military-industrial complex resulted in the century's second great expansion in engineering manpower. Amid this transformation, thousands of engineers in the rank and file worried that they had sacrificed their autonomy to a technological society. Project-based contracts resulted in frequent layoffs, especially as government support declined in the early 1970s, prompting disgruntled engineers to declare themselves "high-class migrant labor."[24]

What was striking about these reformers, radicals, and disaffected sympathizers was the degree of notoriety they attained and the frequency with which they reconciled with technology's critics. The shelves in engineering libraries bear the mark of the profession's introspective turn. At Princeton, where the activist professor Steve Slaby led a campaign against secret research, forgotten volumes written by engineers with titles including *The Mute Engineers* and *Understanding Technology* stand beside worn copies of Ellul and Mumford.[25] When I uncovered reformers' archives, the pattern was similar. Later I found scores of classroom attempts to bridge the "two cultures" divide between engineers and humanists. Then there were the underground journals and newspapers that printed book reviews, poetry, and conversion stories.

Initially I was convinced that following "radical" engineers who called for social responsibility and read "dangerous works" offered the best means to revisit the themes raised by Layton and Florman. A study of engineering's dissidents would shed light on questions of broad import to scholars and to engineers today concerned with their profession's social vision. It would add to recent historical studies of how the upheaval of the 1960s altered the nation's technical communities in a manner that moves beyond arguments of immoral complicity in the military-industrial complex or of activist scientists as moral heroes. In *Disrupting Science*, for example, Kelly Moore shows that dissenting scientists in organizations such as the Union of Concerned Scientists (UCS) and Science for the People (SftP) helped enhance science's authority in public controversies, but as a consequence diminished their individual status as "objective" experts.[26] Jennifer S. Light describes how "defense intellectuals"—RAND Corporation scholars, National Aeronautics and Space Administration (NASA) managers, and Lockheed executives—shared their cybernetic visions with urban politicians in an effort to improve American cities, making strange bedfellows with community activists.[27] And Fred Turner argues that a motley crew of communitarians, military-industrial researchers, and journalists promoted information technology as a countercultural tool, in the process introducing a vision of technology as libratory into the commercial mainstream.[28]

Despite participating in the scientists' movement, the environmental movement, and the counterculture, the stories of engineering's reformers remain largely untold. They have been overlooked in part because the notion of radical engineers is counterintuitive, even contradictory. But engineers have escaped attention also because they are rarely considered to be distinct and worthy cultural subjects when compared to scientists, intellectuals, or artists. It is precisely in their role as producers of culture, however, that the history of engineers is valuable.

Engineers often deride the utility of language, privileging instead scientific objectivity and the ability to get things done. Nonetheless, even the most taciturn of engineers is an intellectual of a sort because engineering is an inherently normative practice. While engineers frequently collapse *is* into *ought* or deny that a sociological imagination shapes their designs, to engineer is to build a social vision into material reality.[29] What better way to understand engineering's normative character than by studying those who made explicit the gulf between the *is* of society as they saw it and the *ought* for which they risked their careers? Once we traverse boundaries between modernity's assumed critics (typically leftist intellectuals) and its proselytizers (businesspeople, scientists, and engineers), we find new means for explaining the cultural and intellectual changes of 1960s America.

Engineers' appropriation of critical theories, finally, illuminates how *technology* became one of the ascendant keywords of the twentieth century and how its meaning is not static. From the 1930s to the present, few concepts have been as central to engineers' visions of self and society as *technology*. For professional elites, technology has served as the organizing principle of a vision that identifies engineering as a vocation with the authority to judge. Among rank-and-file engineers it has structured stories that connect individual labor to a larger social good. Technology initially was given meaning as a powerful but abstract force of social change not by engineers but by social theorists. In early twentieth-century America, Thorstein Veblen, William Fielding Ogburn, and others filled a semantic void in a society of rapid industrial growth employing the term *technology* to describe the mix of scientific rationality and interlocking machines that seemed to be novel agencies in the modernization of society. In bolstering the "ideology of engineering" in the 1930s, progressives and conservatives alike appropriated the concept—casting it as masculine, powerful, complex, scientific, and emancipatory.[30] But there is no such thing as an ideology of engineering in general, and the rhetorical marriage between *engineering* and *technology* is not as natural as it first appears.[31]

Critical theories of technology in the 1960s and early 1970s were powerful because they conceptualized humanity's relationship with its devices in a manner compelling to engineers disenchanted with technology's promises. Concepts appropriated from technology's critics—specifically the belief that technological systems embodied arrangements of power and authority and that those systems could take on a life of their own—changed engineers' notions of responsibility and their possibilities as agents of social change. For these engineers, social progress would be achieved through the search for alternatives to technocratic rationality. In this regard understanding how engineers appropriated alternative visions of technology offers a usable past from which to approach current debates about climate change and globalization, which have roots in the era, are debated in similarly existential terms, and present novel alternatives to what engineers should know and whom they should serve.

I hope my contributions remain those just described, but as I followed reformers and radicals in their campaign to remake engineering, an additional set of actors kept demanding to be heard. Even the most dismissive critics of "intellectual Luddites," I found, countered with social-theoretical ammunition. In many cases they borrowed from the very opponents they rejected. In *Chemical Engineering Progress*, for example, Ralph Landau, the president of petrochemical processing company Halcon International, Inc. and an inductee into the National

Academy of Engineering (NAE), portrayed energy development in a world with "limit on growth" as society's greatest challenge, but he rebuked critics like the activist biologist Barry Commoner for their "socialist" solutions. To Landau such scientific authorities became "amateurs (laymen)" when they reached beyond their fields of expertise. Quoting the conservative political columnist Henry Fairlie, he claimed that "removed from his own discipline no one is more vain than the intellectual." And yet Landau asserted that engineers were in fact intellectuals and that they needed to master the "interplay of technology and politics" because they were best equipped to "anticipate the outcomes." He emphasized pedagogical reform, arguing that engineers required training in economics, business management, and political science, but that the humanities—the greatest source of skepticism—should be eliminated from engineering education. Underlying Landau's contradiction was his awareness that preserving "efficiency and freedom" necessitated making knowledge claims about society.[32] His target audience moreover was not Commoner or Fairlie, but rather the thousands of engineers in the American Institute of Chemical Engineers (AIChE).

Engineers' struggle to restore progressive meaning to technology was not between progressive reformers awakened by critical theory and a conservative majority devoid of new vision, nor was it principally between engineers and anti-technologists; rather, it was in the main a conflict among social theorists within engineering itself. In professional societies, grassroots reformers competed with marquee national directives to define the engineer's responsibilities. In the classroom, instructors and textbooks gave competing explanations for technology out of control, the nature of historical change, and the future careers of their students.

By the time engineers had decided to air their grievances at the 1971 IEEE Convention, technical solutions were rarely offered outside the context of encompassing normative visions of technology. For a dedicated core these were full-fledged political ideologies, presented in social theoretical texts, but for the majority of engineers they were porous worldviews that bolstered self-understanding and political identification by appropriating themes from multiple sources. Either in book-length monographs or letters to the editor, these patchwork worldviews had an epochal tone that contextualized the profession's malaise in reference to the *longue durée* of engineering's past and offered a privileged vantage on the future.

In an occupation with a million members with diffuse bonds, only a small minority were visible participants in debates about out-of-control technology. Engineering's reformers and radicals as well as its corporate and academic leaders were significantly more literate than the rank and filers. Many had advanced degrees, were politically active throughout their lives, and were avid readers beyond their fields of practice. The connections between an engineer of Packard's stature, academic engineers such as Paschkis or Slaby, and a manufacturing engineer for General Motors (GM) may have been tenuous, but the minority's struggle to redefine technology's meaning were formidable in reimagining what counted as engineering. Indeed it was precisely engineering's real and imagined bonds that were at stake.[33] Normative visions of technology were not transferred directly from on high to the factory floor nor from a single leaflet in Columbus Circle, but rank-and-file engineers encountered dominant images of technology and its discontents in professional journals, annual meetings, pedagogical training, television, lunchroom jokes, and a range of other formal and informal media.

As professional elites struggled to claim cultural authority over technology, they too reached out to intellectuals and politicians to explain technology's ills and to offer a path forward. In contrast to the scattered ideological character of the dissident minority, by the late 1960s professional elites reached a consensus view that—building on the work of political theorist Langdon Winner and historian Rosalind Williams—I identify as an *ideology of technological change*.[34] This was way of seeing society posited that rapidly accelerating technological advances had produced an array of negative, unintended consequences that stressed existing institutions and patterns of life ill-equipped to handle them. These problems had only recently emerged, and were the result of technology's nature. Consequently engineers could not be blamed for the country's technological dilemmas. Engineers moreover would be best equipped to adapt society to technology by maximizing its positive opportunities and minimizing its negative effects. As Simon Ramo, a founding member of the NAE, wrote in his 1969 monograph *The Century of Mismatch:* "There is an imbalance, a mismatch between accelerating technological advance on the one hand, and social adjustment and maturing on the other. The appropriate action is to work at speeding up social advance and improving the selection of priorities for technological investment."[35]

It might seem strange to describe this worldview of accelerating change and lagging social adjustment as an ideology; after all, this is the semantic universe we encounter daily. "Science and Technology form a two-headed, unstoppable change agent," the journalist Joel Achenbach professes in a recent *Washington Post* article notable only for the ordinariness of its theme. "We vaguely understand that this stuff is changing our lives. . . . We're just hanging on tight."[36] But such a position is not modernity's folk ideology; it echoes a social theory crafted by academics, journalists, and prominent engineers.

This vision has gone comparatively understudied by historians of technology in favor of analyses of "technological progress" and "technological determinism," categories that cloud the specific origins and uses of how technology is understood in practice. An ideology of technological change is not simply a variant of technological progressivism, and it is more than technological determinism.[37] It is a normative philosophy based on a series of flexible concepts that render the past and present understandable and the future manipulable through expertise. Emphasizing the novelty of its discovery absolves vested interests of past responsibility for technology's deleterious consequences and positions the discoverer as best suited to make decisions about future innovations.

An ideology of technological change had immediately beneficial outcomes for engineers. It restored identifications of engineering as a profession with expert authority, quelled intraprofessional dissent, and collapsed *is* back into *ought* by naturalizing technology in a way that denied legitimacy to "ideological" opponents. Moreover it did not call attention to the fact that tens of thousands of engineers were directly or indirectly employed in weapons development.

By linking engineering to inevitable technological forces, however, this vision has proved hazardous.[38] As was the case for engineers drawn to critical theorists of technology, it was far easier to explain what was wrong with society than it was to implement solutions in a world that failed to succumb to ideals. More significant, an ideology of technological change portrayed engineers as adrift in the historical wave of innovation. It maintained much of the heroic imagery with which engineers

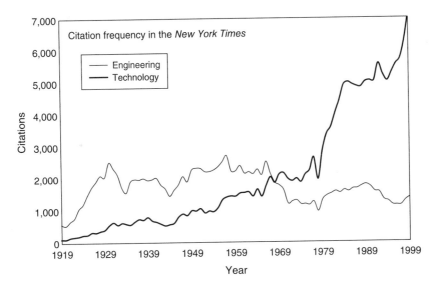

Figure 1.2
Semantic reversal of *engineering* and *technology*

defined themselves in the early twentieth century—positing, for example, that elite engineers could assert control over technology by channeling its autonomous power. But the terms of command had been reversed. As Williams explains in her analysis of MIT's restructuring in the 1990s, "while 'technology' expands its rhetorical reach, that of 'engineering' shrinks."[39]

Reconstructing the *is* and the *ought* of technology from the engineer's pen thus helps bring into focus a dominant vision of technology in our time—one that continues to preoccupy engineering's professional leaders even as they strive to overcome it.[40] Engineers' reconciliation with competing theories of modernity bears on the history of engineering, the cultural-intellectual history of the United States, and the conceptual framework with which we perceive contemporary life. To understand the evolution of technology's cultural meaning we need to explore sweeping social and technical transformations in the postwar era, but we need also to investigate a small group of engineers as they assumed the mantle of intellectuals.

To explain how cultural conflict and intellectual appropriation altered the worldviews of American engineers, I take inspiration from the historian Darrin M. McMahon who has argued that the making of "enlightenment" in eighteenth-century France was as much a product of the bromides and pornographic satire of anti-*philosophes* whose "reactionary" conservatism was cast in modernist language and influenced the terms of modernity itself.[41] In similar fashion, engineers' efforts to make technology "responsible" and "humane" were met by resistance that put forth a competing vision using the same words. The stakes for *philosophes* and anti-*philosophes* were orders of magnitude higher, the battle lines clearer, and the outcomes greater, but it was not accidental that engineers described their predicament akin to "Lavoisier to the guillotine" or lambasted a "chorus of physiocrats." Between 1964 and 1974 a brand of confrontational ideology in American politics created an environment in which it seemed the country was on the brink of revolution.[42] Engineers were hardly culture warriors on par with Barry Goldwater or Abbie Hoffman, much less the Minutemen or Weathermen. Indeed most engineers struggled with the very notion of being "political." Nonetheless, they came to perceive America's crisis of technology with different normative visions that amounted to competing systems of moral understanding.[43]

To bring into view the legacy of this moment when engineers feared they had lost control of technology, I do not recount in depth the history of specific programs of environmental science, risk assessment, engineering ethics, and the like. Nor do I dwell on differences between engineering disciplines—for example, chemical engineers and filtration devices or aerospace engineers and the supersonic transport (SST). While there is much to be learned from such studies, my target is the conceptual framework that informed and connected engineers' labor to social vision. Whether conceived of as "mechanics of the middle class," "trusted workers," "hired servants," or simply as "people," engineers possess technological skills with real power to transform the material world.[44] But, before engineering designs and before political action, there are words.

2

From System Builders to Servants of The System

Usually the engineer's work, important though it may be, is unknown and uncredited outside his own organization and professional group. He is the behind-the-scenes expert, the follow-through man who gives way to the scientist, architect or designer when the bronze plaques are put up. There's a saying in the aerospace industry that: "When a rocket is successful, it's a triumph of science, but when it's a dud it's blamed on a technological failure of engineering."

All of these things the engineer learns to shrug off and go about his job, as his predecessors have done for 5,000 years. If he does his job well he has the deep satisfaction of knowing that he has created something that will last and will make a solid contribution to the common good.

—Theodore Wachs Jr., *Careers in Engineering*[1]

These are the myths that sustained the "shock troops of industrial capitalism."[2] The engineer was a hero once—the professional who grew out of his artisanal origins to become humanity's steward of technological progress, forded rivers, commanded industry, and conquered nature for the benefit of all civilization. Samuel Smiles, the British success-philosopher who instructed industrious souls to pull themselves up by their own bootstraps, once deemed the engineer worthy of his own book of *Lives*.[3] Just as Giorgio Vasari had claimed Michelangelo equal to the statesmen of antiquity, so too stood the railroad pioneer George Stephenson and the mechanic James Watt. The engineer once came to the American frontier to make it compliant for settlement and commerce. Clad in a leather jacket and engineer boots, he (always a he) commanded lesser men, raised dams and factories. Later he donned a laboratory coat to invent new wonders and defend the nation against its enemies. Poised to reconcile society's conflict between labor and the captains of industry, there was a time when a Stanford educated mining engineer became president of the United States. During the Great Depression even as Hoovervilles proved as lasting a legacy of the "Great Engineer" as the dam that bears his name, the engineer once brought electrification to the rural poor. When global war again erupted, the engineer again assured victory.

Somehow this engineer was lost. By the early 1960s popular imagery cast engineers as dark-suited, short-sleeved, nondescript, white men with black-rimmed glasses. College-educated residents of suburbia, they lived in artificial environments and worked in vast teams with narrow responsibilities. Even the profession's public outreach presented an ambivalent picture. In its 1966 volume *The Engineer*, Time Life lauded engineers' societal contributions but emphasized their specialism. Beginning with the segment "12,000 Engineers to Make a Car," it dispelled notions of the "solitary, boot-shod adventurer" in a panorama of identically attired draftsmen:

The fact is the great majority of engineers today work in teams—behind desks or in laboratories—tackling mutual problems with slide rules, computers and microscopes. The modern engineer fits no single mold. He is part scientist, part inventor, part technician, part cost accountant—and almost always a specialist in a narrow field. . . . A chemical engineer may do nothing but study better ways to manufacture quick-drying paints.[4]

The Engineer focused public attention on quandaries that dominated professional societies and national commissions throughout the postwar era. Were engineers designers, scientists, or applicators of existing knowledge? Builders of structures or manipulators of symbols? Professional experts or technical manpower?

A range of interdependent factors influenced this transformation from a virile pioneer to someone who watched paint dry, but none were so fundamental as the relationship between the federal government and the corporations and universities in which engineers were trained and employed. Spurred by mobilization during World War II, the profession underwent structural change matched only by the corporate growth of the early twentieth century. In service to the twin projects of national defense and space exploration, hundreds of thousands of engineers were minted and received significantly more academic training than ever before. New fields of technical knowledge further expanded the range of

Figure 2.1
"12,000 engineers to make a car"
Source: C. C. Furnas and Joe McCarthy, *The Engineer*, Life Science Library (New York: Time, Inc., 1966), 17.

activities encompassed by the term *engineering*, as science and technology became central to the cold war political economy.

This technological prosperity redefined what it meant to be an engineer. Its rewards, however, were unevenly distributed and presented material and existential challenges. Engineers were drawn to the aura of science but struggled to maintain cultural identifications distinct from it. The demographic and conceptual expansion of an already diverse profession loosened the bonds of shared knowledge and practice as specialization and the fluctuations of federal contract-based hiring redefined notions of service and advancement.

As professional elites strategized how to enhance engineering's authority with respect to scientists and corporate and government employers, the rise of environmental, civil rights, and antiwar movements further recast the terms of what it meant to be an engineer. The nation's political polarization challenged the profession's racial and gender homogeneity, its basis in neutral rationality, its contribution to environmental degradation, and its collaboration in weapons-making. As the United States became a technological society characterized by nuclear weapons, digital computers, and chemical pollutants, engineers increasingly were portrayed as duplicitous technocrats. "There is no stopping the engineering mentality," the journalist Gene Marine would write in his 1969 exposé *America the Raped*. "We can only try to stop the Engineers."[5]

By the end of the decade, engineers from the professional elite to the rank and filers interrogated their cultural, economic, and political status. They asked: who was the engineer? How much in control was he, not only in his job, but over the devices he produced? What were the values that underlay his work? And, could a renewed vision of engineering address public anxiety about technology run amok? This chapter explores the origins of—and challenges to—American engineering's progressive ideals to explain why engineers became partisans in debates about technology and the common good.

Envisioning an Engineering Past

Engineering in the United States has been a segmented occupation at least since the late nineteenth century, prompting historical analysts to deem the title *engineer* to be "virtually meaningless."[6] Practitioners' level of formal education has ranged from a high school diploma to the PhD. Engineers have labored as CEOs, managers, rank-and-file employees, subcontractors, consultants, soldiers, civil servants, and university educators in organizations ranging from self-employment to corporations with two hundred thousand employees. By 1960 there were over forty professional engineering societies—some with as many as twenty-five specialized divisions—and almost half of those claiming to be engineers did not belong to one.[7] Clearly, there is no such thing as "engineering culture" in any deep anthropological sense; there are only myriad subcultures.[8]

Nonetheless, despite laboring in differing institutional contexts with varying norms and practices, few ideals have meant more to technical practitioners than their identification as engineers. Evolving conceptions of the engineer have tied individual work to global progress based on material improvement. Shared images of self and society have created solidarity and given meaning to changing work relations through economic depression and war. The images that took root, those that were rejected, and those that remain perpetually contested help explain the existential character of engineers' social thought in the 1960s.

There is some truth behind the frontiersman image that shaped popular conceptions and self-identifications of American engineering in the first half of the twentieth century. By 1850, in addition to a few thousand West Point-trained Army officers, there were roughly five hundred civilian engineers serving a population of twenty-three million.[9] Training was characterized by apprenticeship in the construction of canals and railroads. These "civil engineers" emphasized a shared masculinity with the laborers and tradesmen they managed but stressed their superiority derived from intellect, initiative, and moral respectability. As engineers gained economic and cultural status—the most successful becoming independent consultants—they viewed themselves analogously to doctors and lawyers, valuing autonomy, self-regulation, and social responsibility. To consolidate their status and transfer esoteric knowledge, new technical schools such as Rensselaer Polytechnic Institute (RPI) and MIT were established and the nation's elite colleges adopted the applied sciences to train gentlemen polytechnicians.[10]

The most persistent myths of American engineering, however, took root in an environment in which they bore little relation to common experience. Engineering in the United States developed largely as a consequence of industrial capitalism and its organizational form, the vertically integrated corporation. Between 1880 and 1930 the number of engineers increased from 7,000 to 226,000 to fill the ranks of General Electric (GE), Westinghouse, American Telephone and Telegraph Company (AT&T), Du Pont, Ford, Union Carbide, Monsanto, and Standard Oil.[11] More than a hundred new technical schools opened their doors to the sons of America's rural middle class as science-based industry created fields of chemical, electrical, and metallurgical engineering.[12]

Engineers fashioned identities in corporate America between the poles of science, management, and labor. The heterogeneity of specialized knowledge and social position that characterized engineering in the 1960s already was present at the turn of the century. Independent consultants and a small group of graduate-trained industrial scientists championed professionalism based on autonomy from big business by restricting the definition of *engineer* to those engaged in design or applied science. The first national member organization, the ASCE, defined ethical obligations to clients and the public good. Subsequent societies adopted elements of the ASCE model but were far more accommodating to corporate employers.[13] Indeed most engineers in the member societies considered a position in top management to be the height of professional virtue. Professionalism, they argued, gave engineers authority to extend techniques for controlling nature and machines to the control of men. Both of these models, however, obscured the fact that by the early twentieth century, the majority of technical workers were salaried employees with little likelihood of attaining autonomy or executive responsibility. Rank-and-file engineers might have organized in labor unions, but ideals of scientific mastery and business management reinforced an ethic of individual merit. Skilled technicians were encouraged in the new schools, in corporate culture, and in member societies to embrace a distinct self-image as *engineers*.[14]

Engineering's visionaries sought to elevate the status of their profession by claiming technology to be the wellspring of social progress, which was a direct outcome of their organized creativity.[15] Images of engineering's past, present, and future supported political action and gave meaning to everyday life by fusing notions of middle-class respectability, self-improvement, masculinity, individualism, loyalty, scientific objectivity, and profit-making with material advancement. Where mechanization appeared in manufacturing, where networks for transporting goods were needed, the engineer—the practical scientific expert who could invent ingenious devices and put them to mankind's benefit—was to be found. This flexible normative vision coalesced at the turn of the century and reached its peak in the early days of the Great Depression, gaining credibility not only through circulation among engineers but also through widespread repetition by intellectuals, artists, and popularizers.[16]

The ideology of engineering, however, was contradictory at its core; crafted in exchanges between one set of elites seeking corporate leadership and another trying to make engineering an autonomous profession. Reformers, inspired by political progressives, asserted that the same methods that enhanced machine efficiency and worker productivity could be applied to a rapidly developing society rife with disorder and waste. Calling upon the virtue of social responsibility, they attempted to unify the engineering societies to overcome what they perceived as the corrupting influence of utilities and manufacturers. But the engineer as agent of social progress was equally at home in private industry where responsibility and service were mobilized in a vision of the corporation as an inherently reformist enterprise that generated maximum efficiencies by unlocking engineers' collective potential.[17] Near-universal agreement that engineers were masters of technology—reinforced by innovative corporate public relations campaigns—patched over conflicting notions of service.[18]

The Great Depression, however, revealed the limits of engineers' professional and political aspirations. The breakdown of the global economy seemingly presented reformers with the opportunity to achieve the societal management championed by their ideology. Instead, it strained the tenability of engineers as technocratic experts.[19] Admonitions about repairing the "social machine" devolved into accusations of blame for economic catastrophe, identifying international markets rather than technology or American business as culprits.[20] At the same time high unemployment attracted nearly 10 percent of the nation's engineers to organized labor.[21] To combat unionism, society officers and corporate managers attempted to rally the rank and filers around private industry as a source of technocratic solutions. But the emphasis on sweeping historical vision dissipated as professional societies struggled to support members and remain financially solvent.[22]

Cold War Transformations

World War II dramatically altered the practices and norms of engineering in the United States. Long known as the "physicists' war," behind nearly every science-based weapons systems was an integration of industry, government, and academe, supported by a labor force of tens of thousands of engineers, technicians, and semiskilled workers.[23] The rise of the aeronautics industry, though less heralded than the atom bomb, illustrates key dimensions of what came to be known as the "military-industrial complex." The industry's workforce rose from forty-nine thousand in 1939 to over two million by 1943 as craft manufacturing was replaced by mass production. Automobile manufacturers entered the aeronautics market and their engineers accelerated output and reduced skilled labor by designing automated techniques.[24] As a consequence of industrial conversion, big business grew and diversified such that by 1944 corporations with over ten thousand employees accounted for 31 percent of all American workers. Just ten firms—the burgeoning aircraft makers as well as GM, Ford, Chrysler, Bethlehem Steel, and GE—received 30 percent of all prime government contracts, and the top hundred corporations accounted for three-fourths of the value of US manufacturing.[25]

In the war's immediate aftermath, engineering journals cautiously anticipated another economic downturn; instead, the prosperity forecast by the nation's largest businesses became a reality. The standard of living rose to new heights, measured by the proliferation of consumer technologies. By 1956, 73 percent of American families owned automobiles, 96 percent of homes with electricity had refrigerators, and 86 percent had televisions.[26] Space age design was found everywhere from television casings to breakfast cereal packaging, as social progress and the American way of life seemed an inevitable consequence of modern science and technology.

If the postwar era was a consumer's republic, however, it was undergirded by a war economy. During the Korean War, defense expenditures returned to World War II levels of twenty billion dollars annually, and would climb to eighty billion by 1970. Aerospace firms were the largest beneficiaries; between 1961 and 1967, federal contracts accounted for 88 percent of Lockheed's business.[27] Domestic prosperity and military security were intertwined in justifications of this new world. Corning advertised that its kitchenware was constructed with "an astounding new missile material," which "Science created," while Chrysler boasted that the "missile-making experience" of its electrical engineers translated into better automobiles.[28]

Hundreds of thousands of engineers were designed to serve cold war progress. By 1960 engineering was the most common occupation for white-collar males in the United States. According to the Bureau of Labor Statistics, approximately one out of every fifty men in the labor force identified himself as an engineer. There were eight hundred thousand engineers in America, a nearly fourfold increase from 1940.[29] In Florida, Tennessee, Texas, and especially California—which alone received a third of all federal R&D funding—high-tech industry transformed the population and migration patterns of technical labor.[30] Private industry and utilities employed 71 percent of the nation's engineers, while 15 percent worked for government agencies. However, when government contracts were considered, 40 to 45 percent of American engineers were either directly or indirectly in the government's employ. Of the million engineers in the workforce, approximately 75 percent were employed in just 1 percent of all firms. Corporations with a workforce of ten thousand or more employed 35 percent of America's engineers.[31]

Engineering expanded rapidly not only in size but also in conceptual diversity as higher education became the dominant path of occupational entry.[32] Bolstered by the GI Bill, tens of thousands of bachelor's degrees were awarded yearly. According to the Engineers Joint Council (EJC)—an organization founded in 1941 to promote the collective goals of the member societies—graduate trained engineers, once a tiny minority, expanded to 31 percent by 1969.[33]

As crucial as the increase in educated manpower was the ascent of *engineering science* as the profession's leading paradigm. Engineering science had its roots in curricular changes of the 1920s and 1930s, but the federal government's postwar commitment was its main impetus.[34] As the historian Stuart W. Leslie writes, weapons systems dictated the research and curricular agendas of the nation's universities:

Indeed, those technologies virtually redefined what it meant to be a scientist or an engineer—a knowledge of microwave electronics and radar systems rather than alternating current theory and electric power networks; of ballistic missiles and internal guidance rather than commercial aircraft and instrument landing systems; of nuclear reactors, microwave acoustic-delay lines, and high-powered traveling wave tubes rather than Van de Graaf generators, dielectrics, and X-ray tubes. These new challenges defined what scientists and engineers studied, what they designed and built, where they went to work, and what they did to get there.[35]

MIT and a handful of elite schools—including Johns Hopkins, Princeton, and Stanford—were empowered in the military–industrial–academic nexus. MIT in particular became a symbol of national science and technology policy. Time Life's *The Engineer* dedicated an entire chapter to the Institute to illustrate the challenges faced by America's "new crop" of engineers.[36] In MIT's alumni magazine, *Technology Review*, Gordon Brown, dean of engineering, described the Institute's curricular changes, contending that undergraduate education formed the nucleus of a school surrounded by graduate training and interdisciplinary research centers, which in turn were encompassed within the basic sciences.[37]

The development of intercontinental ballistic missiles (ICBM) and space exploration also elevated a generation of engineers to executive responsibility in the nation's top universities, corporate research laboratories, and electronics and systems firms—the latter of which were founded by entrepreneurs who had left GE, AT&T, and the like. This elite group earned PhDs in the engineering sciences at the California Institute of Technology (Caltech), MIT, Princeton, and Stanford in the 1920s to 1940s. Their experiences during World War II contributed to their faith in the systems approach, of which the ICBM project was the definitive success. At its peak, building the delivery system for America's nuclear arsenal involved seventy thousand employees in twenty-two industries, eighteen thousand scientists and engineers, seventeen contractors, two hundred subcontractors, and two hundred thousand suppliers.[38] In addition to cutting-edge research in the fluid mechanics of atmospheric reentry, this hybrid military–industrial–academic project required new manufacturing capabilities, logistical planning, public relations, and the conviction that social phenomena could be reduced to objects in socio-technical models.

Figure 2.2
A "new crop"
Source: C. C. Furnas and Joe McCarthy, *The Engineer*, Life Science Library
(New York: Time, Inc., 1966), 98.

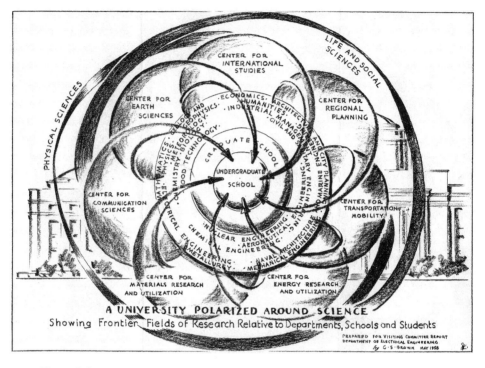

Figure 2.3
MIT's engineering science ideal
Source: Courtesy of MIT Museum.

Status and Anxiety on the Endless Frontier

Along with its myriad opportunities, engineering's structural and epistemic changes generated profound stresses on professional institutions and individual careers. From the perspective of the National Science Foundation (NSF) and the EJC there never were enough engineers to support the arsenal of democracy. Land-grant and independent technical colleges underwent a contentious shift in priorities from practical to scientific training amid cries that engineering education had forsaken its manufacturing mission.[39] The project-based, single-buyer defense economy resulted in frequent job transfers and mass layoffs, while the secrecy of defense work contributed to the fragmentation and compartmentalization of engineering labor.[40] Obsolescence became a watchword among the rank and filers as the knowledge explosion created new subfields and eroded barriers between disciplines.[41]

To systems planners and engineering scientists, this was a golden age in which engineers were positioned to be creative leaders. They nonetheless lamented the backwardness of their profession and its subservience to—rather than direction of—the system. Engineering's movers and shakers became preoccupied with occupational status and public image. When previously booming student enrollments leveled off in the early 1960s, they attributed the slowdown to confusion over what engineers actually did, undue attention to the sciences, and an emphasis on *technology* over *engineers* in popular media.[42] Charting how the NSF distributed its awards, professional leaders developed "a vague sense of insecurity and defensiveness" about their relationship to science.[43] In 1964 the EJC warned that engineers were becoming technical support, "on the sidelines" of policy-making.[44] J. Douglas Brown, dean of the faculty at Princeton, argued that if engineers ignored the vast changes of the scientific age, "professional schizophrenia" between traditional manufacturing ideals and the fluid research economy would cripple engineering's long-term relevance.[45] These expressions of doom and gloom were deployed to garner support for a reinvigorated professional ethos. As Brown wrote, engineers would be best positioned to tackle America's challenges once they enhanced their public image, standards of education, and government service.

To shore up occupational boundaries from below, educators and member societies differentiated engineering from lesser forms of technical labor. Throughout the 1960s they collaborated to establish a bachelor's degree and member society participation as grounds for entry into the "organized engineering profession." At the same time "subprofessional" training in "engineering technology" was institutionalized at two-year vocational schools.[46] Consequently, when the EJC counted in 1964, it identified 678,000 engineers, putting the total without bachelor's degrees at under 15 percent; whereas self-reported census data recorded engineers without degrees at 40 percent.[47]

Enhancing status with respect to science proved more difficult. Reading technical journals from the 1950s and 1960s, one sometimes got the impression that the biggest threat to the United States was its physicists. Even as engineers increasingly worked alongside scientists—not only as university consultants, political advisors, and research directors but also as rank-and-file employees—they felt marginalized.[48] Daniel E. Noble, a prolific inventor for Motorola Inc., placed the blame for engineering's marginal image on the press and on corporate marketing that called engineers "scientists" to enhance company prestige. "Perhaps an engineer is an engineer, is an engineer, is an engineer," Noble wrote, "but in the public mind, which is sometimes either confused, stupid, or imbecilic, he is a vague shadow."[49]

Scientists' own existential doubts fueled engineers' vitriol. As historian David Kaiser shows, cold war R&D blurred occupational categories that challenged ideals of science as an intellectual fraternity. In response, scientists appropriated "the language of suburbanization and mass consumption to express both deep-seated wishes as well as fears."[50] Drawing from contemporary social theory, physicists claimed that demands for technology were eroding pure science. Engineers, for their part, addressed their wishes and fears by casting engineering as science-*plus*. Both scientists and engineers probed the unknown, their argument went, but where scientists could retreat from social relevance, engineers were forced to confront "reality in all its aspects."[51]

In addition to vocational boundary disputes, engineers found themselves entangled in controversies about postwar cultural values. In the 1950s the "organization man" emerged as shorthand for the paucity of individual autonomy and authentic experience in a borderless corporate world. Engineering played a decidedly negative role in the critique. William H. Whyte Jr. in his bestselling *Organization Man* blamed exaggerated engineering manpower shortages for biasing higher education against the humanities.[52] More critically, the sociologist C. Wright Mills identified "the Technician" as the archetype of stultifying white-collar existence. Mills lamented the marginalization of thought as corporate values spread to the intellectual, who like the Technician, was

solidly middle class, a man at a desk, married, with children, living in a respectable suburb, his life a tight little routine, substituting middle-brow and mass culture for direct experience of his life and his world, and, above all, becoming a man with a job in a society where money is supreme.[53]

This diagnosis resonated among engineering educators and corporate managers. Indeed Mills' description was not far off the mark of the profession's evolving self-image. In 1967 John Dustin Kemper's textbook, *The Engineer and His Profession*, identified the character of the engineer in terms informed by Mills:

The engineer generally *is* a clean-cut young man in a white shirt, tie, and business suit . . . a college graduate (and there is increasing likelihood that he has an advanced degree) and is at home with science and mathematics. He is a resident of Suburbia, and is a commuter. . . . He is ambitious (though not feverishly so), generally believes he can advance only by "going into management," and may vaguely resent the alleged "fact."[54]

By the mid 1960s the organization man motif was a staple of corporate recruiting aimed at convincing engineers that their work environment was not "soulless" and that employees would not become "robots." They promoted individual independence by dismissing "restrictive engineering" and showing openness to new cultural norms; one ad declared, for example, that "your beard won't bug us."[55]

Reformers targeted education as a critical site for raising engineering's status. Vannevar Bush—the electrical engineer who famously laid out a vision for the NSF—chastised educators for producing engineers who lacked the flexibility of their scientific peers. "It is to be regretted," he argued in a 1952 speech that was reprinted in engineering journals well into the 1960s, "because we cannot give up the most cherished attributes of our profession—versatility and resourcefulness—without making our jobs unattractively one-sided and conventionally drab."[56] Engineering scientists argued that the master's degree should replace the bachelor's as the first "professional degree" because advanced training created flexible thinkers who could adjust as technology and theory evolved. They also called for a greater emphasis on the humanities and social sciences to train the whole man.

But even before future engineers could be trained properly, schools needed to attract the right sort of students. Career literature aimed at high schools presented engineers as rocket builders to convey the profession's excitement and societal importance. These texts reflected tensions between organizational needs and desires for national leadership. They only occasionally presented heroic profiles of systems executives and instead romanticized the experiences of the technical masses laboring to put Americans on the moon. The book *Careers in Engineering*, for example, gave a human face to manpower shortages through the tale of a NASA recruiter searching for "first-rate engineering minds." The demand for talent was urgent, *Careers* explained, because the Soviets led two to one in the training of engineers. Descriptions of the space program transitioned into an account of how "advanced weapons of war" required

Will Olin turn you into an organization robot?

These soulless creatures who do and think only what they're told are an anathema to Olin. We want humans. The best we can get. We want people who think. And say what they think. People who can still dream. And wonder. And get mad when things foul-up.

Specifically engineers, chemists and business majors to work in chemicals, aluminum, packaging material, sporting arms and ammunition, brass, paper and energy systems in 60 locations throughout the country.

For more information, see your Placement Officer or write Mr. Monte H. Jacoby, College Relations Officer, Olin, 120 Long Ridge Rd., Stamford, Conn. 06904. **Olin**

The best thing we have to offer you is you.

Olin is a Plan for Progress Company and an equal opportunity employer (M & F).

Figure 2.4
"Will Olin turn you into an organization robot?"
Source: Olin Corporation, "Will Olin turn you into an organization robot?"
Journal of College Placement 30, no. 1 (October/November 1969): 83.

specialists from every technical branch. Even traditional fields like civil engineering contributed by designing attack-proof launch sites. To explain how engineers simultaneously served the nation and private enterprise, *Careers* posited a hypothetical manufacturer of ceramic materials, "Spaceage Products, Inc." Spaceage's research chemist discovered a heat-resistant material transparent to radio waves, which a team of engineers subsequently developed for "nose cones on guided missiles" but also for use in cooking utensils.[57] Organization and the team were core virtues, representing unity, service, restrained pride, and obedience. Another career guide, *So You Want to Be an Engineer* relayed engineering's rewards but prepared its initiates for anonymity: "The ones really responsible for the first moon landing will probably never be named in the history books. The credit will go to the crewmen and to their country—but this first moon landing, when it finally comes, will really be a triumph of engineering."[58]

Efforts to improve education and public image were tied to top management's ambitions of shaping national policy. In 1964 a core of systems entrepreneurs, research directors, and corporate executives—almost all of whom had served in the National Defense Research Committee during World War II—led the charge to formulate the National Academy of Engineering, an elite body with twenty-five initial members, to advise government with a voice distinct from the National Academy of Sciences (NAS). J. Herbert Hollomon, the Department of Commerce's first assistant secretary for science and technology and one of the NAE's initial visionaries, argued that engineering's image and support problems were of its own making—engineers shunned public service, lacked social responsibility, and thus had no lobbying influence on a Congress of technical laymen.[59]

Not all engineers in management, however, considered the influx of government capital or the embrace of engineering science to be an unmitigated boon. A stream of editorials in *Chemical Engineering Progress* lamented "big government," "big research," "government, government, government," and the "government/industry myth." The "heavy, heavy influence" of the government was strongest in engineering education, they argued, which had "strayed far afield" of its industrial mission in the pursuit of esoteric scientific research.[60] Those not immediately involved in aerospace and defense R&D described the sectors as essentially "nonproductive" and contended that more engineering man-power went into the design of a new automobile than a missile.[61] Indus-trial managers claimed that engineers were better suited to learn on the job than in graduate training, arguing that advanced education for a tiny elite should not disrupt the flow of "garden variety" engineers for service in "less romantic" manufacturing jobs.[62]

Disagreements about who engineers ought to be—scientists or indus-trialists, entrepreneurs or company men—were amplified by tensions between expert authority and bureaucracy that resurfaced as elite reform-ers with political aspirations called upon engineers' professional obliga-tions. Much of the discussion, however, was not about public service or autonomy but rather "treating professionals professionally" by not making engineers punch time clocks. Management was generally happy to support continuing education and professional society membership because the "feeling of being professionals" was a retention strategy and a deterrent against collective bargaining.[63]

Proponents of what came to be known as the "new professionalism" warned against limiting engineers' obligations to company service and the technical state of the art. In *The Engineer and His Profession*, Kemper contended that what united the million individuals who called themselves engineers was

basic devotion . . . to *technical advancement*, and to a large degree, this has also become the devotion of the general public. Some might even say that technical advancement has become engineering's religion, in the broad meaning of the word: namely, a set of principles and beliefs to which a man devotes his life.[64]

Kemper argued, however, that by itself this devotion was neither profes-sional nor sustainable because it was not anchored in the belief that "*human welfare* is the ultimate justification for technology's existence."[65] If engineers lost sight of their public responsibility, he prophesized, they would face a societal uprising.

Unraveling Prometheus

Predicting rebellion when in the midst of one did not require exceptional clairvoyance. By the time Kemper's textbook reached students, the Civil Rights Act and the Great Society had been matched by race riots, political assassinations, student insurgents, and hippies trying to levitate the Pentagon. Underlying the political violence were increasingly extremist theories about why the United States was a "sick society," with a diverse and disjointed new Left characterizing contemporary life as authoritarian and an emboldened Right railing against the breakdown of the social order.[66]

Technology, "the engineer's religion," emerged as a principal fault line in the nation's culture wars, channeling cold war anxieties about status and identity into an existential conflict over the meaning of progress. Critiques of technology called into question the ends of engineering service and the limits of technical solutions. When the space program achieved its goal of putting a man on the moon, for example, even the usually roseate *Time Magazine* claimed that "if Apollo was a victory for US engineering genius, it could not disguise American failures at home."[67]

By the late 1960s professional leaders acknowledged the erosion of engineering's public image and warned of the consequences. In the early 1960s, as technologies central to engineers' self-fashioning and economic security encountered suspicion, the society journals pushed back against what they interpreted as muckraking crusaders. After a 1967 request from the EJC, technical journals published scores of articles on technology & society topics.[68]

When criticisms of specific devices and systems shifted into a broader interrogation of the nature of technology, engineers interpreted it as a wholesale assault on their identity. In defenses aimed as much at engineers as the broader public, employers sought to demonstrate their mastery over technology's unwanted impacts. Recruitment advertising cast critics as obstructionists and stressed engineers' contributions to social progress. In 1967, for example, the Westinghouse Electric Corporation offered a "modern fable with technical overtones" in which a college graduate named Jack wanted to do something important with his life. "He could [have gone] on to grad school, or join[ed] the Central Intelligence Agency (CIA), or volunteer[ed] for social welfare service, or participat[ed] in a protest movement," but he chose engineering instead. At Westinghouse, Jack met Jill, another young engineer, and joined a nuclear power project in an "underdeveloped nation in a faraway land." On the job, he captured a "subversive agent," completed the plant on time, and proposed to Jill at the airport. The text was interspersed with images of a national award for his exploits, a dark-suited recruiter with black glasses, and political buttons that read "PEACE" and "ANTI-PROTEST."[69] Other companies offered a *mea culpa* for technology's ill effects and outlined steps for reform. The Ford Motor Company ceded: "We're one of the causes of air pollution. (We're also one of the prime solutions)" while Pacific Telephone called for a "Truce" that began, "You don't like us. Sometimes we don't like ourselves."[70] More common, however, were full-throated defenses of corporate innovation as the answer to pollution and urban poverty. Westinghouse again was a pioneer: "After the statesmen, politicians, educators, journalists, clergymen, sociologists, and theorists have finished talking about what should be done, we turn it over to the engineers to do it. Nothing, from the pyramids to Apollo 16, would have been built without them. Nothing important tomorrow will be built without you. So who needs engineers? We do."[71]

Member societies reached out to students to dispel notions that they were hostile to the concerns of youth. The National Society of Professional Engineers (NSPE), for instance, dedicated the October 1968 issue of its journal to student authors. Princeton senior William W. Hill advocated that engineers serve the urban poor. "We are not a group of subversive socialists who see as the answer to the ills of society the overthrowing of big business and the capitalist system," he wrote, "Quite the opposite is true."[72]

Major steps also were taken to interest women and minorities into the profession. At a symposium cosponsored by the Society of Women Engineers and the Engineering Foundation, Stanley W. Burriss, president of Lockheed Missiles and Space Company, outlined the profession's "pragmatic" appeal to diversity:

As engineers and scientists we are all painfully aware that in the short-term—we hope—the word "technology" has become synonymous with pollution and war. Our young people are no longer impressed with man-on-the-moon accomplishment. The fact that our problems require more and better technology has failed to penetrate the din of rock and roll or whatever piper is predominant at the moment. This near-term pseudo idealism will result in a long-term problem—namely, a shortage of technologists.[73]

The polarization of American culture that Burris lamented, however, had emerged on a smaller scale within engineering itself.

The widening gulf between conceptions of technology, progress, and engineering was most dramatic in the same institutions that previously represented cold war progress. At MIT, the student magazine *Tech*

Go Westinghouse, Young Man! *A modern fable with technical overtones.*

Once upon a time there was a young senior in college named Jack who couldn't decide about his future.

He wanted to do something worthwhile after graduation.

But there were so many things to do, it was hard to decide. He could go on to graduate school, or join the social welfare service, or participate in a protest movement . . . or he could enter the business world.

Many of Jack's friends urged him to steer clear of big industry.

"There are no challenges in air-conditioned offices," they warned.

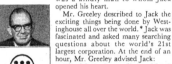

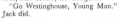

And it was a challenge Jack wanted — the kind of challenge his forefathers faced on the frontiers.

Then he met a Mr. Greeley.

Mr. Greeley recruited college students for Westinghouse Electric Corporation. He was a kindly man to whom Jack opened his heart.

Mr. Greeley described to Jack the exciting things being done by Westinghouse all over the world.* Jack was fascinated and asked many searching questions about the world's 21st largest corporation. At the end of an hour, Mr. Greeley advised Jack:

"Go Westinghouse, Young Man." Jack did.

The first few weeks were difficult. There was so much to learn.

Jack was to discover that at Westinghouse, learning was a way of life, that a career with Westinghouse was one long process of education and re-education.

Later Jack was permitted to decide which of six big groups he would like to join.** Jack selected the Westinghouse Electric Utility Group.

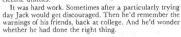

With the Electric Utility Group Jack learned about water processing, about power generation, about underground distribution, and many other things. Jack had not realized how important to the survival of modern man is the world of electric utilities.

It was hard work. Sometimes after a particularly trying day Jack would get discouraged. Then he'd remember the warnings of his friends, back at college. And he'd wonder whether he had done the right thing.

Then came Jill. Pretty, intelligent, warmhearted Jill. Jack had met Jill at the drinking fountain in the Utility Group Water Province Department.

Jill was an engineer with Westinghouse (Editor's Note: Women are welcome at Westinghouse, an equal opportunity employer).

Although the work became more and more difficult and the hours longer, Jack with Jill at his side persevered.

Then came an assignment to join a team of Westinghouse engineers and scientists. The team was being sent to an underdeveloped nation in a faraway land to help rebuild a large coastal city.

Jack and Jill's assignment: Help build a power plant that would use nuclear fuel. (Nuclear fuel lasts longer than coal or oil. And it's cleaner.) Energy from the nuclear plant was used to change salt water from the nearby sea into fresh water that the poor people of this country could use as drinking water.

Working late one evening on the job site, Jack caught someone in the act of sabotaging the construction of an extra-high-voltage distribution system. This system would bring power from the nuclear plant hundreds of miles into the inland areas of the country.

After a dramatic chase through the winding streets of the city, a chase in which the international police and CIA participated, Jack captured the subversive agent. A grateful nation presented him with its highest award.

Finally, the project was completed. It was hard work but it was good work. Thanks to the Westinghouse team, millions of people would live better.

The citizens of the country were grateful. They wanted Jack and Jill and the others to stay . . . offered them more than their present salaries as an inducement . . . but Westinghouse fringe benefits more than offset this offer.

At the airport, where a sad but affectionate crowd of citizens gathered to see them off, Jack turned to Jill and asked:

"Will you marry me?"

Jill smiled and said: "I will if you promise to let me join you on other equally important turnkey projects that Westinghouse is coordinating in some of the major cities in the United States."

Jack promised, and they lived happily ever after.

Moral: Awaiting you at Westinghouse are challenges, hard work, building block education, adventure, some travel and, yes, even romance.

You can be sure if it's Westinghouse Ⓦ

For further information, please contact: L. H. Noggle Westinghouse Educational Center, Pittsburgh, Pa. 15221.

*UNDERSEA EXPLORATION   MASS TRANSIT   WATER DESALTING   AEROSPACE TRAVEL   AUTOMATED PARKING GARAGES   PROGRAMMED LEARNING   TOTAL-ELECTRIC CITIES

**CONSUMER PRODUCTS   INDUSTRIAL·   CONSTRUCTION   ELECTRONIC COMPONENTS & SPECIALTY PRODUCTS   ATOMIC, DEFENSE AND SPACE·   ELECTRIC UTILITY

Figure 2.5
"Go Westinghouse Young Man!"
Source: Westinghouse Electric Corporation, "Go Westinghouse, Young Man!" *Engineering and Science* 30, no. 4 (January 1967): 1.

Engineering News described the formation of a civil rights group in 1965 as the arrival of the "Dissident Organization."[74] Later, on a spring night in 1967, the countercultural hero Timothy Leary—sitting in the lotus position on an oriental rug—debated the uses of lysergic acid diethylamide (LSD) with professor Jerome Lettvin—who walked commandingly around the stage wearing horn-rimmed glasses, a short-sleeve white shirt, and tie.[75] Then, in March 1969, students and faculty shut down the Institute for a teach-in on its role in the military-industrial complex.

A minority of engineers challenged colleagues to face their collaboration in weapons development. These reformers embodied heterogeneous motives and organizational strategies that extended beyond professionalism traditionally defined. Interest-group politics, unionism, and religiously motivated opposition all garnered constituencies. The "closed world" of the defense industry was a key site of dissent. For the rank and file to voice discontent was a high-risk proposition that has left only a small imprint on the historical record. In his memoir *Blue Sky Dream*, however, the journalist David Beers provides a glimpse into aerospace industry disillusionment. He recounts how his father, an engineer at Lockheed, experienced a breakdown of idealism. At first, writes Beers, the segregation of corporate life kept his family and their community ignorant of the unrest around them. Inspired by the advice of a colleague, however, Beers's father amassed readings that questioned the values of American society and its technology. Critical of the moon landing as a cover for military contracts, the elder Beers wore a black armband to work in support of the 1969 "March Against Death." An hour after a coworker asked "who died?" and he responded, "about fifty thousand Americans and untold Vietnamese," he found himself in the office of a supervisor who informed him that such an action "was the sort of thing that could get a man's 'ticket pulled.'" Beers removed the armband.[76]

In engineering cultures, souring attitudes about technology were closely associated with fluctuations in the defense economy. In 1971 the NSF reported unemployment of 3 percent among engineers—well below the national average, but the highest since the Depression. In the space program and defense industries unemployment was far worse, with the official jobless rate reaching 9 percent in Seattle and 7.4 percent in Orange County.[77] Independent surveys, however, contended that the true unemployment rate was over 20 percent nationally.[78] Tales of aerospace engineers turned taxi drivers put a human face on an equal percentage of engineers now employed outside the technology sector.

Economic recession in the high-tech labor market exacerbated partisanship among engineers. Congress's decision in 1971 to terminate the SST was the boiling point. Trade journals savaged Wisconsin senator William Proxmire, who had led the charge to eliminate the two billion dollar program, by railing against the "pseudo-liberals" and "professional bleeding hearts" who opposed the program on the ground that the money might be better spent on social programs.[79] One inventive letter to the editor expressed engineers' frustration in verse:

There once was a Welfare State
That engaged in extended debate.
All things technological (not ecological)
Suffered a dire fate.
Then one day when the wheels stopped spinning
And the lights went out for good,
They called for a man to fix them
And found that nobody could.[80]

Odes to the Senate, however, were matched with challenges that the "rhetoric of free enterprise" concealed dependency on government contracts and layoffs were the price of misguided priorities. "Advancing technology is certainly important to any world leader," one Navy lieutenant responded regarding the SST, "but I feel, as many do, that our government and industry have been emphasizing hardware technology at the expense of humanitarian technology."[81]

As segments of the engineering profession debated the nature of humane technology, their worldviews could be rife with inconsistencies. In *The Mute Engineers*, an anonymously penned analysis of his profession, for example, a New Jersey engineer debunked twenty-one "myths" of engineering, including the assumptions that the engineer is a scientist, that as a professional he has control over his decisions, that the engineering societies represent their members, that engineers have a positive self image, that more engineers are inherently better for the country, and that engineers are "a dull, ethnocentric and irresponsible lot."[82] Citizens were revolting against "rampant technology," Anonymous contended, because the engineer was an abject failure:

He suffers the incognition of a profession where he has neither the prestige of the scientist nor the bargaining power of the blue collar worker. He is harassed by the manager, snubbed by the scientist and continuously solicited to by the blue collar union. He is lost. *He does not have an identity.*[83]

In its iconoclasm, however, *The Mute Engineers* fell back upon engineering's most enduring mythology. What distinguished the engineer was "his rugged individualism, his aspirations to an higher ethics, his education and training, and most important, his position as the progenitor of technology in a technological age." Anonymous argued that protests by humanists and the young could "create a climate for change," but engineers would be the vanguard because "only technology can cure." Disrespected and abused, the engineer was unleashing "a cry in the dark," not of despair but "of a newborn, of a new way of life, for itself and for the world."[84]

Conclusion

This then is how the engineer was lost. From the perspective of a professional elite, engineers were ceding leadership to physicists, sociologists, and politicians. More shockingly, as technology opened new vistas of humanity, antiprogress intellectuals assailed the engineer as an "all-American whipping boy."[85] Yes, a civilizational clash with the Soviet Union for technical supremacy dominated the content of engineering labor. Yes, America's cities were in a deplorable state. Yes, technology generated ill effects. But in their desire for change, critics disparaged the only source of true progress. This is "The Test," wrote Arthur Kantrowitz, director of the Avco Everett Research Laboratory. "The revolution brought about by the magnificent union of science and technology is now threatened by a massive counterattack driven . . . by the very real and justified fears of the consequences of technology but perhaps still more by the unpredictability and the present lack of control of technology."[86]

At the same time, a vocal minority of engineering reformers and radicals—educators, disgruntled employees, and dropouts—stressed the contradictions between professional myths and lived experience. Their nation was engaged in a war utilizing the tools they had built. The space program had become a political and economic well run dry. Decades of indiscriminate dumping and industrial accidents raised an environmental fervor. "Ecologically speaking the engineer has been one of the most immoral people in history," a recent graduate declared at an EJC symposium on "Moral Issues in a Technological Society." To be sure, he continued, engineering was one of the "most successful social revolutionary forces in history." But, in their greed, naivety, and apathy, engineers had abdicated their duty to social progress. Only through reinvigorated public service would they again be "true professionals."[87]

Unemployment, specialization, and worker alienation contributed to this revolt of the engineers. So too did jurisdictional questions about safety standards, pollution abatement, and energy policy. Few events were as polarizing as the Vietnam War. But to attribute the surge in engineers' political energy to economic interest or a range of individual technical problems misses the extent to which engineers saw themselves as participants in a contest about the nature of social progress. Beset by boundary problems that were loosening their substantive and rhetorical command of technology, reformers in executive authority and in the rank and file set out to remake engineering as the progressive force of its imagined past.

3

Technics-Out-of-Control as a Theme in Engineering Thought

Technique has been extended geographically so that it covers the whole earth. It is evolving with a rapidity disconcerting not only to the man in the street but to the technician himself.
—Jacques Ellul, *The Technological Society*[1]

On every page an engineer turned during America's unraveling there was a palpable sense that *change* was in the air. The rank-and-file Boeing electronics designer seeking order behind the televised collage of race riots, psychedelia, and a giant leap for humankind might have found it in *Newsweek's* Independence Day issue of 1970. The "extreme crisis of confidence," historian Arthur Schlesinger Jr., wrote, is a result of "the incessant and irreversible increase in the rate of social change" caused by technological advances that "make, dissolve, rebuild and enlarge our environment every week."[2] His manager's faith in objective rationality could be confirmed in the sociologist Daniel Bell's still widely read *End of Ideology*. "The mass society," Bell argued, "is the product of change—and is itself change. It is the bringing of the 'masses' into a society from which they were once excluded" that rendered critique into little more than "an ideology of romantic protest against contemporary life."[3] The engineering student seeking a more distinctly ideological understanding of change turned to the Port Huron Statement, in which the Students for a Democratic Society (SDS) juxtaposed material affluence and nuclear proliferation against stifling social conformity. "Without new vision," the SDS pleaded, "the failure to achieve our potentialities will spell the inability of our society to endure in a world of obvious, crying needs and rapid change."[4] For a radicalized engineering minority, no one identified the crisis of contemporary existence as well as Lewis Mumford, who, in the first installment of his *Myth of the Machine*, wrote: "Never since the Pyramid Age have such vast physical changes been consummated in so short a time. All these changes have, in turn, produced alterations to the human personality, while still more radical transformations, if this process continues unabated and uncorrected, loom ahead."[5]

If partisans across the spectrum held at least one belief in common, it was that technology was at the center of the revolution in which citizens of advanced industrial societies found themselves. It is difficult to overstate the profusion of ink spilled about the nature of technology in the 1960s. Aided by increases in federal funding for social scientific research and an expansion in paperback publishing, a new genre of technology & society writing emerged that cut across academic disciplines, political positions, and popular audiences.[6] Its texts deployed a host of neologisms—"megamachine," "technostructure," "technetronic era," "technique"—to conjure contradictory images of unstoppable global-historical forces and free will, creativity, and responsible choice.[7]

In their efforts to restore progressive worth to their profession, engineers became consumers and sometime-producers of the technology & society genre. Radicals, reformers, and defenders of the status quo alike discussed and debated Hannah Arendt, Kenneth Boulding, Pierre Teilhard de Chardin, the Club of Rome, Barry Commoner, René Dubos, Peter Drucker, Jacques Ellul, Paul Ehrlich, Victor Ferkiss, R. Buckminster Fuller, John Kenneth Galbraith, Paul Goodman, Jürgen Habermas, Robert Hutchins, Ivan Illich, Arthur Koestler, Melvin Kranzberg, Herbert Marcuse, Gene Marine, John McDermott, Marshall McLuhan, Lewis Mumford, George Orwell, Ayn Rand, Theodore Roszak, Charles Reich, C. P. Snow, Oswald Spengler, Irving Stone, Paul Tillich, Alvin Toffler, Thorstein Veblen, and Lynn White Jr.

This chapter brings into relief the competing ideologies with which the engineering profession's visionaries sought to reform their vocation. It does so by tracing engineers' appropriations from and contributions to the technology & society genre. Technology & society texts were not the proximate cause of engineering reform or radicalism, but they became important resources for confronting long-standing assumptions about what it means to be an engineer. Exploring the networks in which engineers' ideas about technology & society circulated and the intellectual practices of engineers who drew upon those ideas provides a conceptual map with which to interpret the diverse landscape of engineering reform in the 1960s. Drawing on the insights of the political philosopher Langdon Winner, I begin with a broad overview of the technology & society literature, emphasizing the critical intellectuals that so roiled engineers. I then explain why a minority found the worldview of the critics so appealing. Finally, I describe how social scientific policy makers and engineers in top management collaborated in shaping a different vision of out-of-control technology.

Technology as Ideology

Looked at with forty years' distance, the explosion of critical inquiry about technology can be credited to a combination of factors, including efforts to interpret the social meaning of atomic power and industrial automation, a backlash against highly publicized environmental disasters, the growth of federal bureaucracy, opposition to the Vietnam War, neo-Marxist disillusionment with Soviet communism, and a broad recognition of what historian Thomas P. Hughes called humanity's "literally making a new material world."[8]

Technology & society literature was produced by a comparably heterogeneous group of commentators. The first significant critiques came from scientists in the anti-nuclear movement of the late 1950s and early 1960s.[9] Shortly thereafter, a nascent environmental movement called attention to the hazardous effects of pesticides.[10] Marvels of the consumer economy also came under suspicion when Ralph Nader assailed automakers for knowingly selling dangerous products.[11] Ambivalence about technology extended to the civil rights movement, with Martin Luther King Jr. chastising a society that produced "guided missiles and misguided men."[12]

By the mid-1960s exposés about specific technologies shifted to critiques of "technology" at large. These works came in the form of transatlantic imports such as Ellul and Marcuse as well as indigenous cultural criticism. Interdisciplinary seminars at Columbia, Harvard, MIT, and less prestigious universities were important sites of conceptual development. Government agencies likewise devoted manpower and printing offices to the problems of out-of-control technology. Foundations and think tanks from the Center for the Study of Democratic Institutions to the RAND

Corporation identified the social study of technology as central to their missions. Multinational firms conducted in-house seminars and funded academic projects. Finally, Charles Reich and other academic commentators suggested that the counterculture articulated new philosophies of technology expressed through lifestyle rather than monographs.[13]

In his landmark retrospective of the era's technopolitical visions, *Autonomous Technology*, Winner demonstrated that in this crowded marketplace of ideas, visions of technology clustered around two poles with "almost no middle ground of rational discourse."[14] One was a utilitarian-pluralism that interpreted technology's ill effects as inherent to its evolution. Pesticides eradicated malaria but reaped havoc on ecosystems; highways enhanced mobility but generated suburban sprawl; the birth control pill aided family planning but sparked a redefinition of sexual mores; the green revolution reduced famine but fueled a population explosion. In passing, Winner referred to this point of view as an "ideology of technological change."[15] Its advocates argued that to bring technology's societal disruptions under control required a balance of technocratic management, political representation, and social adjustment.

The second pole, which Winner called a "theory of technological politics," was a "set of pathologies" that claimed modernity's evolving systems were foreclosing the possibilities of humane existence. He identified nine recurrent themes of technological politics:

Artificiality   Mankind has built a "second creation"
Extension   The system's range of capabilities is now global
Rationality   Life in the system is logically ordered
Size and concentration   The system requires massive capital commitment
Division   Labor has become specialized and isolated
Complex interaction   The system is unintelligible to users and designers
Dependence/interdependence   Humans lack autonomy and power is shared unequally
The center   Control of the system takes place at a core of power
Apraxia   Small failures can throw the entire system into chaos[16]

As Winner characterized it, the "central hypothesis" of technological politics was that once the world constructed by men was sufficiently advanced, it would continue to reconstruct itself through technology, becoming more difficult to control, much less abandon. Proponents of this perspective asserted that technologies ultimately were the material embodiment of political philosophies. Because of technology's systemic nature, they contended, reform could not focus on individual devices, but rather needed to confront the socio-technical structure of society writ large.[17]

Winner (who appears in chapter 7 as an actor rather than an analyst) studied discourses of technology as an advocate for technological politics as a viable moral philosophy. Consequently he treated a "theory" of technological politics and an "ideology" of technological change asymmetrically. He also was more concerned with the prospects of these visions than he was in tracing the networks in which they were produced. It is important, however, to stress that technological politics was not something that "happened" to engineers. Professional engineering cultures sustained a robust debate about the meanings of technology & society as active readers and intellectuals in their own right. Reconstructing how engineers came to view technology as being out of control demonstrates the dialectical formation of these competing ideologies. Most important, it forces us to take seriously how the ideology of technological change achieved cultural supremacy.

Engineers as Critical Theorists of Modernity

In the mid-1960s themes of technological politics trickled into engineering journals, society meetings, and even design methods. Engineers drawn to critical theories of technology typically were academics or defense industry employees. Texts in this vein appealed because they explained the formation and consequences of the "military-industrial complex" that controlled the economy, created engineering unemployment, and bore responsibility for the war in Vietnam. Engineer reformers rarely took critical social texts as definitive societal models. Though they agreed with criticisms of technology's neutrality and their profession's lack of autonomy, they generally rejected the totalizing character of the critique.

Three theorists—Ellul, Mumford, and (to a lesser extent) Marcuse—provided the greatest sense of "disappointment, anxiety, even menace" among engineers.[18] Each stressed technology's dominance in contemporary life, connecting material changes to a totalizing rationality that threatened individual autonomy and human creativity. Ellul, who was a leader of the French resistance during World War II, offered an eschatological worldview of systematic technology. He defined "*technique*" as "the *totality of methods rationally arrived at and having absolute efficiency* . . . in *every* field of human activity."[19] Mumford presented a similar analysis of the system. He argued that since humanity's origins, technology had existed in competing forms of "democratic" technics (small-scale tool-making that involved personal autonomy) and "authoritarian" technics (power and domination achieved through institutional regulation and large-scale organization). Beginning in the sixteenth century, he contended, authoritarian technics had taken on an internal momentum. If society continued in its present direction, the potential for any alternatives would crumble as the logic of technology became absolute.[20] Marcuse likewise identified technology as not simply "the sum total of mere instruments which can be isolated from their social and political effects," but rather as "a system which determines *a priori* the product of the apparatus as well as the operations of servicing and extending it." This system had resulted from an "initial *choice* between historical alternatives," however, once the path of industrial rationality had been made, it curtailed all other possible futures.[21]

For theorists of technological politics, changing the system would come through dialectical analysis of society's *is* and *ought*. Mumford argued that technology could serve spiritual and artistic fulfillment if reshaped in accordance with the "human center." "Plainly," he concluded, "it is not the mechanical or electronic products as such that intelligent minds question, but the system that produces them without constant reference to human needs and without sensitive rectification when these needs are not satisfied."[22] Marcuse had also suggested that technology would play as important role in a "transcendent" society as it had toward domination in the existing system.[23] Still, these critics left the steps toward building a society based on the human center undefined. In the absence of concrete solutions, they argued, the first task was to become aware of one's place in the system.

Technology's critics pointed not only to the problems inherent in technical systems, but also to the need for technical experts to develop philosophical introspection. The liberal economist John Kenneth Galbraith, for example, called for responsible professionals to reorient the "technostructure," as they were the only ones with expertise to understand how the system functioned and with the power to make technology serve human needs.[24] The law professor Charles Reich likewise argued that "what we need is education that will enable us to make use of technology, control it and give it direction, cause it to serve values with which we have chosen."[25] They may have been surprised to find a technical readership that agreed.

Reading critical social theory offered disaffected engineers clarity and intellectual authority. Consider, for example, the manner in which the ideas presented in Ellul's *Technological Society* found their way into the engineering press. Ellul claimed that technique was a precondition of science, that it rendered technicians into components in a system they did not control, and that, as it propagated, humans lost their souls. Writing in the *Journal of Engineering Education*, John M. Boyd, a mechanical engineer at the Worcester Polytechnic Institute, embraced a similar perspective in his essay "Science is Dead—Long Live Technology!" He offered an enthusiastic primer on Ellul in a schema that characterized technology as "rational and automatic, artificial, centralistic, universal, autonomous, self-augmenting, amoral, [and] monistic." Boyd struggled to find human agency in its development:

[T]he role of man is simply to participate in the sorting operation. He does not control the operation, for the process is set by the nature of technology. As an analogy it has been said that in a technological culture man represents the slug in the slot machine. He activates the mechanism, but he does not participate in the process.

Despite this self-augmenting autonomy, Boyd believed engineers had a responsibility "to teach how man can use technology rather than be used by it." To do so, he encouraged peers to engage with "sincere critics," among whom he included Dubos, Mumford, and Reich. He admitted that these sources began from negativity, but he claimed that a critical interrogation was necessary to make engineering serve individuals rather than the system.[26]

Other engineers incorporated themes of technological politics into the design process. James D. Horgan, a biomedical engineer at Marquette University, for example, created an analytical technique called the "circle of action" that was based on the *is* of present technology and the *ought* of what a society wants it to be. Horgan quoted Arendt and Marcuse at length to demonstrate (in Marcuse's words) that the traditional notion of the "neutrality" of technology can no longer be maintained. Technology as such cannot be isolated from the use to which it is put; the technological society is a system of domination which operates already in the concept and construction of techniques.

Despite highlighting the "domination" of the present technological society, Horgan sought to convince engineers that the "pessimists" were compatible with engineers' inherently "optimistic view of man." He argued that technology and culture shaped each other reciprocally and that human values were an essential consideration in any technical problem. Selecting goals after reflective interrogation of one's assumptions, comparing those goals to existing conditions, and acting upon them would bring technology closer to human needs. Design was a community responsibility, but engineers had a special obligation to consider

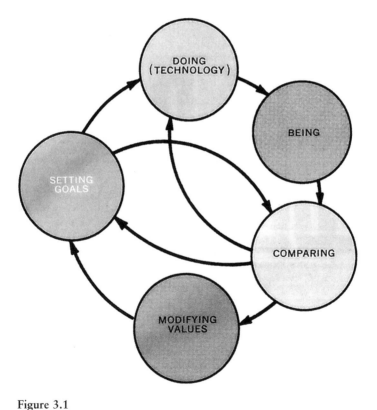

Figure 3.1
Circle of action
Source: J. D. Horgan, "Technology and Human Values: The 'Circle of Action,'" *Mechanical Engineering* 95, no. 8 (August 1973), 20. Courtesy of ASME.

technology's social effects. In all aspects of their lives, engineers could serve human needs by means "partly political, partly technological, and partly a return to a life style that recognizes man's passion to create, to love, to play."[27]

Cases where engineers used technological critiques to formulate their own theories of social action were the exception rather than the rule. More than anything else, theories of technological politics served as organizational aids and as grounds for collective discussion among concerned professionals. Rank-and-file engineers and industrial scientists published alternative newsletters—including Pacific Telephone and Telegraph's *AT&T Express*, Standard Oil's *Stranded Oiler*, General Electric's *GE Resistor*, and Bolt, Beranek and Newman's (BBN) *Signal/Noise*. In this underground press, technology & society literature provided points of reference and solidarity. The first issue of *Signal/Noise* described the origin of an "underground" at BBN. In the winter of 1968–1969, a group of engineers formed with the purpose of "reading, studying, and discussing" Marcuse's *One Dimensional Man*; participation grew and the meetings expanded to "a general critique of society and of the life and work of technical workers."[28] A near universal concern in these papers was unemployment attributed to the economic effects of the Vietnam War.[29]

On the whole, engineers' appropriation of technological politics was unidirectional. That is, engineers encountered ideas in this vein through the published writings of intellectuals. Engineers' appropriation of an ideology of technological politics, however, should not be underestimated. Reconciliation with technology's critics introduced many of the effects associated with ideological conversion.[30] They challenged conceptions of what it meant to be an engineer. They created an awareness of one's subjecthood in "the System," which engendered individual autonomy. They provided a specialized language that contributed an imagined solidarity. Finally, because disgruntled engineers introduced themes of technological politics in professional venues, it was not enough for their leadership simply to dismiss critics as antitechnology cranks.

Ideologues of Change

Indeed, talking about a revolution in the 1960s extended well beyond self-styled radicals, as college deans, professional officers, and corporate executives began to work themes of out-of-control technology into their own visions of reform. In virtually every engineering forum from the mid-1960s onward, one encountered a remarkably consistent set of keywords. The compound gerund *technological change* wove a variety of metaphors from the physical sciences and systems analysis into a coherent historical philosophy. Technological change was *high-velocity, accelerating, rapidly increasing, exponential*, and *revolutionary*. Engineers needed to *keep pace* with the Soviet Union, the knowledge explosion, and technology itself. Achieving social progress required interdisciplinary teams of experts to help the nontechnical elements of society *adapt* and *adjust* by *minimizing* technology's negative *unintended consequences* and *maximizing* its positive *opportunities*.

The rhetoric of inevitable technological change and adaptation was not altogether novel in the 1960s, though it was different in scale and kind. Its core mechanism dated to the origin of *technology* as a keyword in the American cultural grammar. In the 1920s the sociologist William Fielding Ogburn combined anthropological notions of adaptation, macroeconomics, and human relations to argue that invention was the central factor of social evolution. For Ogburn, societal strife was the result of "culture lag" in which "rapid change in one part of . . . culture requires readjustments through other changes in the various correlated parts of culture."[31] During the Great Depression, in *You and Machines*, an American Council on Education pamphlet drawn from the study, he described "The Machine" as a combination of "invention, science, tools, collections of big tools such as are found in factories, machines that move us about, and houses, furniture, etc."[32] As director of President Herbert Hoover's national survey *Recent Social Trends*, "The Machine" became "technology," with Ogburn substituting "technological progress," "ever expanding technology," "material culture," "inventions," and "machines" interchangeably. All of these were "revolutionary," "rapid," and "advancing" and all created "uncertainty," "unanticipated effects," and the "technologically unemployed."[33]

Ogburn's theory of culture lag remained an important concept in postwar social scientific theory and was reworked in discourses of *modernization*, the *scientific revolution*, and *automation*.[34] Until the 1950s modernization theorists had primarily seen the problem of acceleration as one of "pre-modern" cultures, but the rhetorical gap between the first and third worlds closed after Sputnik.[35] In domestic science policy, rapid societal change was attributed primarily to organized inquiry rather than autonomous technology; theorists like Harvard political scientist Don K. Price nonetheless argued that control of the scientific enterprise had become complex and illusory.[36]

The most heated debate about the uncertainties of a technological future centered on industrial automation. On the one hand, anxieties about automation extended decades old debates about Fordism, Taylorism, and technological unemployment with added concerns about computerized control.[37] In 1954, for example, Norbert Wiener's *Human Use of Human Beings* presented an apprehensive vision of the new cybernetic relationship between man and machine.[38] On the other hand, John Diebold, whose book *Automation* popularized the term, contended that managing automation's social impacts would ease human adjustment.[39] Similarly Marshall McLuhan argued that the shift from the centralized mechanical age to the decentralized electronic age interlinked every member of society and provided means for a global village of "ultimate harmony."[40]

By the mid-1960s, these intersecting lines of "postindustrial" inquiry were seen as constitutive elements of the need for a general analysis of technological change.[41] From the mid-1960s into the early 1970s, a cast of academics, futurists, politicians, and engineering managers co-constructed a way of seeing society that accepted technology as neither a neutral tool nor an unmitigated social good. Giving voice to the leading advocates of this distinctive ideology of technological change highlights its mutual construction with an ideology of technological politics, while at the same time reveals the crucial role engineers played in its formation.

Naturalizing Out-of-Control Technology

In 1964 the International Business Machines Corporation (IBM) announced an award of $5 million to Harvard University for a ten-year study of the social effects of technology. The grant followed a free-ranging speech by IBM president Thomas J. Watson Jr. (as well as Harvard's decision to continue with IBM over Control Data for its computing needs), who argued that it was time for "those responsible for technological change [to] have some responsibility also to gauge its consequences for society."[42] The new Harvard University Program on Technology and Society was self-consciously normative. Projects would "identify and analyze the primary and second order impacts and effects of technological change on the economy, business, government, society, and individuals and to suggest possible action programs to anticipate, control, and adjust to such effects."[43]

Deans Carl Kaysen (Public Administration), Don K. Price (Kennedy School of Government), and Harvey Brooks (Engineering and Applied Sciences) appointed Emmanuel G. Mesthene, a relatively unknown RAND analyst, to serve as the Harvard Program's executive director.[44] Mesthene, who had just received his PhD in philosophy from Columbia, argued that the Program would "raise the level of the discourse" about technology and "provide the basis for intelligent action." It would pursue the "big questions" to determine "what the current concern over technology and society all adds up to in the end."[45]

In its first three years, the Harvard Program focused on the relationship between technology and business through projects that investigated market structure, industrial organization, and top management. It suggested, for example, that firms could act as an "alternative mechanism" to local governments and interest groups in urban environments by introducing the systems approach.[46] In 1968, prompted by a critical review from IBM demanding studies of "the negative impact of technology on the individual," the Harvard Program shifted toward technology critics and the nature of social change.[47]

At the end of its fourth year, the Harvard Program released synthetic conclusions about the nature of technology to great fanfare. In January 1969, it earned a front-page headline in the *New York Times* with the provocative title, "Study Terms Technology a Boon to Individualism."[48] In the journal *Technology and Culture*, Mesthene criticized existing theories of technology as polemical. On one extreme were the "military leaders and aerospace industrialists" who understood technology as the "motor of all progress," and on the other were the "artists, literary commentators, popular social critics, and existentialist philosophers" who declared technology to be an "unmitigated curse." According to Mesthene, the Harvard Program was best suited to "understand and control technology to good social purpose" because it stood on objective middle ground.[49]

Mesthene clarified his position on technology's autonomy in the journal *Science*. His article "How Technology Will Shape the Future" replaced the Marxist determinism of linear progress with *"probabilistic determinism."* Technology shaped society by curtailing certain practices and institutions and by providing opportunities for the creation of new ones. It did so directly and indirectly. "First-order" effects of technological change were the initial multiplication and diversification of "material possibilities." Its "second-order" effects were threats to traditional institutions and the adjustments needed to assimilate society to new technology. Mesthene denied technology was the sole factor in historical change, but asserted its primacy over all other factors. A critical effect of technology was its "high probability" for altering individual and social values. He defined values as abstractions of choices. Because choices were always limited (like dishes on a menu), new technologies provided the chance for amenable values to flourish and others to be destroyed altogether with "punishing traumata of adjustment that it would be immoral to ignore." By anticipating the impact of technology on values, however, society could align its values to "remain adequate to human experience."[50]

Behind the Harvard Program—advising it and communicating its findings—was a Who's Who of the nation's power elite. Its steering committee included members of all three branches of the federal government, leaders of academia, organized labor, and science popularizers including: Congressman Emilio Q. Daddario (D-Connecticut); Senator Gordon Allott (R-Colorado); Justice Arthur J. Goldberg; Charles J. Hitch, assistant secretary of defense and future president of the University of California System; Donald Hornig, director of the White House's Office of Science and Technology; Lee DuBridge, president of Caltech; Jerome Wiesner, dean of science at MIT; Robert K. Merton, professor of sociology at Columbia; Walter P. Reuther, president of the United Auto Workers of America; and Gerard Piel, the publisher of *Scientific American*. The largest concentration of interested parties, however, was made up of industry presidents, directors, and chairmen—from Bell Laboratories, the Columbia Broadcasting System, Continental Oil, Cummins Engine, Eastman Kodak, General Dynamics, Lockheed Missiles and Space, Polaroid, and Xerox—who were anxious about potential federal regulations. This bipartisan multi-sector representation reflected the entanglement of interests supported by an ideology of technological change.

Largely because of its backers, the Harvard Program was the most prominent voice in debates about technology in the 1960s. It sponsored the research of 109 economists, sociologists, legal scholars, political scientists, and management theorists including luminaries such as the Nobel Prize winning economists Kenneth J. Arrow and Robert M. Solow. Fourteen courses were taught in connection with the Program and its publications were required reading in over 200 colleges. Projects resulted in 15 books and 165 articles in journals, magazines, and newspapers from *Science* to the *Bulletin of the Atomic Sciences, Business Week, National Elementary Principal, Theology Today, New York Times, Playboy, Saturday Review*, and *Scientific American*.[51] It sent 4,000 copies of its yearly reports to universities, government offices, and corporations. Irene Taviss, head of the Program's Information Center, supplemented the reports with primers on technology & society literature that included review essays, abstracts, and suggested solutions. Its correspondence network also was massive, with contacts from universities, religious organizations, local municipalities, business executives, and rank-and-file engineers.[52]

The Harvard Program's biggest impact was on the federal government.[53] When Congress waded into debates about out-of-control technology, it turned to Harvard for expertise. In 1969, largely due to the efforts of Congressman Daddario, the Committee on Science and Astronautics of the US House of Representatives convened to discuss technology assessment. Brooks chaired the panel, which included testimony from Harvard Program scholars. The summary report—perhaps the only one in the Congressional Record to discuss Ellul's *Technological Society*—concluded that in the very source of contemporary social problems were its solutions.[54] To "discipline technological progress" Congress approved the Office of Technology Assessment (OTA), an organization that would provide expert analysis for nearly a quarter-century until it was eliminated in 1995. During his testimony Mesthene reiterated the positive effects of technology. He described criticism of technology itself as one of the problems assessment might solve. He suggested that if Congress approved a Technology Hazards Board, it also needed a Technology Benefits Board to disseminate positive information via television and other mass media to "allay the antitechnology spirit that is abroad in the land."[55]

The Socio-Technologist

Composed principally of academic social scientists, the Harvard Program was overseen by executives of the nation's largest corporations and became a favored authority of engineering managers. But its "social studies" approach was formed across the boundaries of engineering culture. At the same time that the Harvard Program was spreading its message, public intellectuals within engineering were reaching a similar conclusion that put the task of controlling technology in the hands of a new kind of elite socio-technologist. Later chapters explore in depth how a variety of reform movements interpreted the engineer's relationship to technological change through the frame of the socio-technologist. To provide an initial vantage, we need first investigate the most exuberant advocate of this professional ideal.

Simon Ramo, co-founder of Thompson Ramo Wooldridge, Inc. (TRW) and scientific director of the American ICBM program, was one of the most influential engineers of the postwar era. While running a leading high-technology corporation of the space age, he also was a prolific orator and author. Between 1957 and 1973 Ramo delivered or published nearly one hundred speeches and articles and wrote two monographs on the social consequences of technology.

Ramo's rise to prominence rivaled the Victorian success stories chronicled by Samuel Smiles. Born in Salt Lake City in 1913, he was the son of Russian-Jewish merchants who had migrated west. A skilled violinist, Ramo funded his degree in electrical engineering at the University of Utah on music scholarships, prize money, and weekends in the family general store. He then sped through graduate work at Caltech where he earned his PhD in electrical engineering and physics at the age of twenty-three. In the depths of the Great Depression, Ramo took a job at GE. By the age of thirty, he had acquired twenty-five patents and co-authored with his protégé John Whinnery *Fields and Waves in Modern Radio,* one of the most widely read engineering textbooks of the twentieth century.

Consulting trips to MIT's Radiation Laboratory during World War II opened Ramo's eyes to government-directed research. Convinced that the Manhattan Project had unleashed a "technological Pandora's box," he predicted the need for a continental air defense system. He was anxious to get back to California but had little interest in the factory-like environment of Los Angeles' aeronautics industry, which he described as an engineering "cattle yard."[56] In 1946, however, he accepted an offer from a different sort of airplane maker. Flush with cash and a free reign, Ramo built the Hughes Aircraft Company—nominally overseen by the reclusive billionaire Howard Hughes—into an R&D breeding ground and a magnet for military contracts in aerospace electronics.

When Hughes's eccentricities became a hindrance to securing contracts, Ramo and his friend Dean Wooldridge resigned to form the Ramo-Wooldridge Corporation. Shortly after its incorporation, in 1953, the company was granted lead responsibility for developing the nation's ICBM arsenal, making Ramo a household name as one of America's missile men.[57] Adept at forecasting socio-technical trends, he turned the company (which after a merger became TRW) into a billion dollar enterprise. In 1959, TRW received the first NASA spacecraft contract. Three of Ramo's vice presidents went on to head NASA, and Ramo himself chaired the CIA's evaluation of the race to the moon.

Ramo's success was due not only to his technical and managerial skill but also his vision of the systems engineer as a unique societal agent. In the late 1950s, Ramo championed systems engineering as a novel set of techniques for an integrated world of business, government, and technology. In a series of articles and speeches, he developed a social philosophy anchored in technical experience, altering the language of a "missile gap" and a "space race" into a larger global-historical contest between "systems engineering versus the rapidly increasing complexity of our growingly technological civilization."[58] In 1961, Ramo gave a keynote address at the University of California, Los Angeles (UCLA) on the "Coming Technological Society," which was reprinted in popular, technical, and business publications. Describing a litany of anticipated technological benefits, he nonetheless warned of future domestic and international conflict. "The real bottleneck to progress, to a safe, orderly, and happy transition to the coming technological age," he argued, "lies in the severe disparity between scientific and sociological advance."[59]

As charges against the technological society mounted, Ramo became a public intellectual. Between 1963 and 1973, he reworked his theme of socio-technical mismatch in popular media from *Fortune* to the *New York Times, Los Angeles Times, Cleveland Plain Dealer,* and *Women's Wear Daily.* Additionally he conveyed the message to a diverse network of community leaders and politicians. In 1965, for example, he delivered a keynote address at a conference on *Theonetics* (the study of "God's presence in man's accelerating inter-disciplinary discoveries").

Figure 3.2
The engineer as public intellectual
Source: University Synagogue, *Change and the Human Dilemma* (Los Angeles: University Synagogue, 1963).

His work as a speaker and author was secondary to his executive responsibilities, but they indirectly promoted TRW and helped forge networks with academia and government. In 1967, for example, as part of Caltech's development campaign, he ghostwrote a speech for California governor Ronald Reagan that called for using "advancing science and technology to the fullest . . . to minimize—if not eliminate—the negatives resulting from the high rate of scientific and technological change."[60]

Ramo had an especially influential voice in the engineering profession. He wrote guest editorials and feature articles on technology & society in *Experimental Mechanics, Science News, IEEE Spectrum, Astronautics and Aeronautics, Electronic News, Armed Forces Chemical Journal*, and the *Journal of Engineering Education*. As a founding member of the National Academy of Engineering and trustee of Caltech, the California State Colleges, and the Case Institute of Technology, he also propagated his worldview through personal and professional networks.

While Ramo's position never wavered, the context in which he delivered it shifted dramatically. Faced with a "great antitechnology wave," he argued that the nation needed to avoid paralysis by rising above the obstructionists. Two books from 1969 and 1970 synthesized his social theory of technology. The first, *Cure for Chaos: Fresh Solutions to Social Problems through the Systems Approach*, was an apologia for systems engineering. He argued that the problems of mismatched technical and social change were undeniable, and that many of the critiques of existing socio-technology were justified. But in two sectors—free enterprise and national security—change had been managed with spectacular results and that similar successes could be achieved in "civil systems."[61] The second, *Century of Mismatch: How Logical Man Can Reshape His Illogical Technological Society*, laid out Ramo's larger philosophy of history. He debunked the backward-looking myths presented by technology's critics and focused on predicting what future paths were inevitable and where humankind had choices.[62]

Ramo's ideal technological society was a hybrid of government regulation and free enterprise innovation. In the current system, engineering was directed toward short-term goals to create shareholder profits, producing a situation in which no one was responsible for technology's unintended effects. According to Ramo, crippling regulation was the response favored by technocrats and antitechnologists alike. In between a "robot society" and an impossible "rolling back the clock," however, there was a place for rational management.[63] Pointing to smog in Los Angeles as an example, he argued that no solution had been achieved because (1) the city was a collection of municipalities that could not inspect every car on its roads; (2) a mandate from a single state, even one as large as California, would not induce automakers to change their products; and, (3) the free market would not support the start-up costs of mass transit. Only federal law designed in tandem with the auto industry, transportation authorities, and systems engineers could spur innovative solutions.[64]

Adequate socio-technical education was the biggest hurdle to the flourishing of democracy and capitalism. Ramo argued that a false distinction between technical and social knowledge generated a two cultures divide that worked against solutions. By producing only specialists, engineering educators were losing their race with technological change. To broaden its scope, engineering training needed to include applied sociology and psychology to develop techniques of anticipation. To empower socio-technologists, every American citizen needed to learn about technology's social implications because an educated public would be more "appreciative," better able to express its desires, and better able to weigh technology's trade-offs.[65]

In addition to their cadence of *acceleration* and *mismatch*, Ramo's writings about technology and society had common features. First, not once in thousands of pages of prose did he cite anyone. Beyond skewering the "antitechnological wave" there is no hint Ramo had read any theorists of technological politics. Even intellectuals who shared his perspective went unattributed.[66] Ramo's social texts also were filled with contradictions. He began *Century of Mismatch* by claiming that science and technology were "mere tools for civilized man" but then compared artificially intelligent computers to alien "invaders" and claimed that technological change had become "the controlling factor in altering our lives."[67]

It would be a mistake to conclude that this seeming lack of the rudiments of scholarship from engineering's most public intellectual highlights the amateurism of engineers acting as social theorists. The absence of attribution had multiple benefits. It naturalized the mismatched rates of technological and social change as a self-evident reality, suggesting that the only controversy was about which policies were best suited to close the gap. Ramo's ability to speak and write off the cuff contributed to his status as a visionary technologist. When *The Nation* reported on MIT's 1969 research strike, it described the dissenting scientists with curiosity. Its report closed by referencing a speech by Ramo on mismatched technology that had no direct links to the protests on the ground. *The Nation's* conclusion was that "when a scientist-engineer of his prominence criticizes the country's technology, something significant is beginning to stir."[68]

A Vision of Collaborative Utility

As a native engineering philosophy, the ideology of technological change granted engineers a sense of authority over the muddleheaded thinking of humanists. Whereas engineers drawn to an ideology of technological politics were explicit in their borrowing from intellectuals, many professional elites saw humanists and social scientists as impinging on their jurisdiction. Despite this bravado, when explaining the consequences of accelerating technology, most still reached out to humanists and social scientists. In May 1970, for example, while the National Guard occupied Kent State, the American Institute of Aeronautics and Astronautics (AIAA) invited prominent scholars from the American Academy of Political and Social Science to gather for a "Meet the Social Scientist" event.[69] In another symposium on "Technology and Man's Future," sponsored by the NSF, the electrical engineer Richard C. Dorf attracted an audience of a thousand students, faculty, engineers, and managers. Keynote speakers included Harvey Cox, a Harvard Program contributor and Divinity School professor, and Augustus Kinzel, former vice president of Union Carbide and founding president of the NAE.[70]

The new Society for the History of Technology (SHOT), and its journal *Technology and Culture*, was a critical site of interaction.[71] SHOT could be an engineer's first point of contact with the ideas of Mumford and Ellul, but it more likely served to reinforce an ideology of technological change. Ogburn, who died in 1958, had been slated to be SHOT's first president. Melvin Kranzberg, SHOT's motive force, was a proponent of the culture lag thesis, a defender against technology's critical theorists, and a major engineering education reformer. *Technology and Culture* also was a welcoming outlet for the Harvard Program. In 1969 it printed a special issue to coincide with the release of the Program's *Fourth Annual Report* with a summary article by Mesthene and four commentators, including Ramo, who declared the Program's work to be excellent.[72]

The most common means by which engineers constructed their sociotechnical visions was through the reading and interpretation of published texts. Speeches and articles in professional journals were peppered with technology & society references. The Harvard Program's publications were among the most widely cited. In a 1970 keynote address, Eric A. Walker, president of the NAE, used the Program's *Fourth Annual Report* to dismiss claims that engineering had failed as a profession:

The report says, among other things, that most of the undesirable by-products of technology "are with us in large measure because it has not been anybody's explicit business to foresee and anticipate them." In other words, our basic problem is that no one can be held responsible for the problems of the environment since no one is to blame when everyone is to blame.[73]

Walker recounted American engineering's contributions to progress from the early republic to the modern superhighway. "The problem with all of this," he concluded, "is that while engineers and scientists were working on this Frankenstein's monster we call progress, no one stopped to question how big the monster would be when it grew to maturity."[74]

Under Walker's direction the NAE was anxious to establish its credibility as a federal advisor for taming the monster. When Daddario introduced his idea for a technology assessment agency, Congress requested reports from the National Academy of Sciences and the NAE. While the NAS investigated the political and organizational conditions required for accurate assessment, the NAE used case studies to optimize the process. Headed by cost–benefit pioneer Chauncey Starr, forty engineers in the upper echelons of their profession studied subsonic aircraft noise, multiphasic health screening, and automated teaching aids. Initially hesitant about government intervention, the report argued that assessment could "place in context both the warnings from the prophets of doom and the promises of the enthusiasts."[75] The report emphasized the role of corporate management and systems analysis. Experts from impacted industries needed to serve on assessment boards because they were the only people trained to see technology's long-term effects. The NAE report drew sharp distinctions between "creative" and "defensive" interventions. It was imperative to pursue the "visions of opportunity" afforded by creative assessments and not to overemphasize "problem areas." Echoing Mesthene, the NAE suggested public understanding campaigns would help the citizenry "accept constructive technological change."[76]

Strategies for the management of change extended beyond the NAE into member societies, business periodicals, and new technology assessment journals. In 1969, for example, John R. Moore, president of the Aerospace and Systems Group of North American Rockwell, argued that learning to manage change was the "price of civilization's progress and survival."[77] He called for an extension of the techniques developed in the country's missile and space programs coupled with an awareness of the new organizational flexibility of the high-tech economy. Consulting engineer Walter H. Kohl similarly surveyed the enthusiasm and breadth of the assessment process for managing the "uncontrolled application of technology." Chronicling over forty books, journals, and executive seminars, he argued that a new "interdiscipline" was emerging to manage socio-technical complexity.[78] This was a sentiment shared by the presidents and chairman of the nation's technology industries, who gave management lectures on "Technology—Mainspring of History," "The Limits of Technology," and "The New Entrepreneur."[79]

In this vision of accelerating technology, however, the engineer was not necessarily the agent of command. Harvey Brooks argued in *IEEE Spectrum* that the engineer's position with respect to technological change was the most fraught of all citizens. The problem with technology was not that it was too pervasive, but rather whether it would be possible to "generate technological change fast enough to meet the expectations and demands that technology itself has generated." To do so demanded unprecedented "adaptability."[80] Brooks' vision of the ideal engineer was of a flexible, scientifically trained professional, whose knowledge and goals constantly shifted to technology's demands rather than to society's desires.

Implicit in much of the change talk delivered to the rank and filers was a defense of the status quo against the more radical critiques of technological politics that also doubled as a justification for the system's detrimental effects on individual careers. For lack of a better characterization this was the "responsible employee" variant of an ideology of technological change. While top management and social scientists resolved the socio-technical imbalance, the rank and filers should accept their duty to produce technological change, constantly retooling their own skills lest they become obsolete. In a very early formulation Howard K. Nason, general manager of Monsanto's Research and Engineering Division, quoted Peter Drucker's claim that a postmodern era would require "the acceptance of change—irreversible change—as the normal situation." Its benefits would be spectacular but it would put a heavy burden on individual managers and engineers. "Grant us the wisdom," Nason wrote, "to remove ourselves from key positions when critical introspection reveals that we are no longer capable of such adaptation."[81] In a more epochal rendering, Allen F. Rhodes, vice president at ACF Industries and ASME president in 1971, admitted that everyone, including engineers, experienced alienation and hostility in the "modern super-tribe of an industrial society" but that it was impossible to turn back.[82]

Conclusion

Debates about technology in the 1960s and early 1970s were about concrete problems—the use of high-tech weaponry against an occupied country, the wisdom of dedicating billions of dollars to space exploration, the need for new energy sources—but broad ideological frames structured the proposed solutions offered by both engineers and intellectuals. Beyond epithets of "technocrat" and "Luddite," a core of partisans developed normative visions of technology with differing explanations for out-of-control technology.

Among intellectuals there was an open battle between ideologues of technological politics and technological change. The Harvard Program was a lightning rod in this conflict. In 1969, shortly after the release of its *Fourth Annual Report*, neo-Marxist critic John McDermott issued a harsh appraisal in his *New York Review of Books* (*NYRB*) essay "Technology: the Opiate of the Intellectuals." McDermott called the Program's philosophy "abstract" and "sanitary." He argued that Mesthene was claiming "the cure for technology's problems, whether positive or negative, is still more technology." Pointing to bombing decisions in Vietnam made by computers that chose targets without human intervention, McDermott contended that in Mesthene's terms this was capitalizing on technology's "positive opportunities." Applied domestically, he concluded, this logic was eroding the democratic ethos and creative energies of the majority of humankind.[83]

The Harvard Program similarly directed much of its work toward refuting technology's critical theorists. In his 1967 essay "Technology and Wisdom," Mesthene argued that to interfere with technology's acceleration was not only irresponsible; it was inhuman. "'Stop,'" he wrote, "is the last desperate cry of the man who abandons man because he is defeated by the responsibility of being human."[84] The Program also provided thumbnail sketches of the work of Ellul, Mumford, and Marcuse for executives and government officials so that managerial leaders would not have to read them.[85] To assuage concerns about the influence of technology's critics, the Program conducted a survey that measured popular attitudes about technology. It found optimism was highest among the professional-managerial group, whose high education level should have made them "more susceptible to current critiques of technology."[86] The Program's insatiable optimism was brought to a halt, however, when Harvard terminated it two years ahead of schedule. The historian of science George Basalla penned a caustic eulogy, arguing that it failed to achieve "an important role in taming, shaping, or challenging, on intellectual grounds, the ideas put forth by the dissidents."[87]

Indeed the intellectual dissidents appeared to have the last word. For Mumford, the underlying assumptions of an ideology of technological change were insidious. In the second volume of his *Myth of the Machine*, he claimed that, from this perspective, "to resist change or to retard it in any way was to 'go against nature'—and ultimately to endanger man by defying the Sun God and denying his commands."[88] Writing with the benefit of historical distance, Winner called Mesthene's position "sufficiently young to offer spark to tired arguments, sufficiently critical of the status quo to seem almost risqué. But since it accepts the major premises and disposition of traditional liberal politics, it is entirely safe."[89]

In engineering cultures, the causal and temporal divide between "dissident" and "establishment" visions of technology were not so easily disentangled. Reform efforts were bound up in professional organizations, familial relations, work experiences, and social networks in which themes from the technology & society genre circulated in bastardizations and fragments. Because the notion of autonomous technology spanned the ideological divide, it was not uncommon for engineers to believe Mesthene and Mumford were allies.[90] Most important, while these competing visions offer clarity to the historical analyst, they were not actors' categories. Engineers instead defined their profession's challenges through perennial questions of responsibility, service, creativity, and the proper training of future engineers.

It was obvious to engineers, however, that there were significant disagreements over how their profession should confront its problems. On the one hand, dissidents like Roy Rulseh, a certified professional engineer from Milwaukee, Wisconsin, were skeptical of the "new breed" of engineer-manager and railed against "the impenetrable, stone-headed thinking that causes the utter frustration in college students and slum dwellers."[91] On the other hand, in his inaugural speech as ASCE president, Samuel S. Baxter used the concept of culture lag to placate professional unrest. He admitted that the ASCE was slow to embrace "new and intricate relationships of engineering with the behavioral, social and life sciences," but defended "mature" engineers over the younger dissenters. He argued that engineers could become "middlemen" between those "seeking more development" and those "demanding preservation of the ecology and of the nation's social fabric."[92]

As engineers worked to redefine the relationship between *engineering* and *technology*, many of the terms in their conversation were boundary words—flexible terms that provided the commonalities that make debate possible. Shared concepts were defined and redefined, clouding ideological positions. Despite a spectrum of political beliefs and the frequent appropriation of language between supporters of competing views, however, ideologies of technological politics and technological change were oppositional ways of knowing that embodied oppositional interpretations of historical agency. It was in their conceptual push and pull that engineers across multiple venues of professional life struggled to locate their socio-technical future.

4

The Crisis of Technology as a Crisis of Responsibility

[Engineers] will be responsible at all times for being alert to and informed about undesirable consequences, which could be brought about by the scientific or technological business activities or plans in which they are directly or indirectly involved. These consequences may be of a safety, social, cultural, environmental or economic nature, and of long or short range.

—ASME *Technology & Society Division*[1]

In the history of American engineering few images have been as central to the creation of shared values across institutional and intellectual differences as that of the responsible professional. During the expansion of technical labor in the early twentieth century, engineers crafted public identifications that stressed their unique agency for industrial technology. The Machine Age could not exist without creative intelligence, but the engineer was above all a moralist. Responsibility was the faculty of judgment that distinguished him from the skilled laborer, the scientist, and the businessperson as he brought social progress through efficiency and invention.

Nowhere was this social philosophy trumpeted louder than in the member societies, which informed technical workers how to think of themselves as engineers. These channels for publishing transactions and standards also were the profession's closest analogue to a national public sphere in which vocational ideals were debated and disseminated. Throughout the first half of the twentieth century, the nation's innovators held forth on engineers' achievements at annual meetings and set their philosophy into print in society journals.

But if responsibility has been central to professional ideals, it also has marked fault lines of engineering identity. Based on a nineteenth-century model of individual practitioners working for individual clients, notions of professional responsibility spread in a corporate milieu that generated conflicting lines of duty to technical peers, to science, to capital, to one's employer, to one's immediate boss, and to a broader public. Indeed the member societies that championed the engineer's virtue were compromise institutions that balanced tensions between corporate and professional interests when defining entry standards and instilling norms. Society officers, selected from the ranks of top management, imputed a vision of service that linked the factory floor to the executive boardroom. Technocratic reformers alternatively imagined the societies as a vehicle for developing civic protectors. Engineers in the rank and file forged contacts at meetings and in letters to the editor offered a diverse swath of moral commentary. However, lacking mechanisms of arbitration between employee and employer or even for upholding basic ethical codes, the rhetoric of responsibility in member societies elided conflict by portraying harmonious service to many masters at once.

The crisis of technology in the 1960s was a profound threat to engineers' identifications as responsible professionals. Activists from New Left, civil rights, and environmental movements cultivated a discourse of social responsibility to assert their moral authority for change. Adopting C. W. Mills's contention that a "power elite" governed a system of "organized irresponsibility," many described the nation's technologists either as duplicitous or as captive organization men.[2] Charges of irresponsibility, however, were a lesser threat than those of autonomous technology. How could engineers claim command over technology if they were beholden to its destabilizing forces?

In this chapter, I explore how engineers asserted their place at the center of debates about technology by delimiting responsibilities for its consequences. Throughout the late 1960s and early 1970s the member societies became contested sites in which reformers and defenders of the status quo alike posited a "new professionalism." For a cohort of reformers, social responsibility based on a new ethic of technology was necessary to expand engineering's public service. Society officers likewise advocated for public engagement while warning of the "crusading moralist."[3] Executive managers for the nation's largest corporations also called upon engineers' responsibilities for urban renewal and pollution abatement to rebut critics, mollify employees, and enter new markets. A

National Society of Professional Engineers campaign titled "Are YOU a concerned engineer?" successfully captured the overlapping interest of these competing movements: "We, the engineering professionals, have either got to assume a leadership role in the major environmental and other basically technological issues of our time, or we are going to be clobbered."[4]

Reformers in engineering's largest social institutions, in short, interpreted the nation's technological malaise as a crisis of professional responsibility. While the member societies differed in composition and engagement, they shared a general pattern of response. In the wake of Sputnik, society officers ramped up identifications of engineers as responsible professionals in a desire to maintain unity amid rapid growth, specialization, and patronage contests with scientists. This provided reformers with the opportunity to challenge their organizations to make good on that rhetoric. By the late 1960s cultural criticism further destabilized professional norms and fueled rank-and-file dissent. The largest societies then worked to integrate social consciousness into their missions. Reformers created technology & society divisions, established Washington offices, rewrote ethical codes, and expended hundreds of thousands of words to reconcile ideals of service and mastery with autonomous technology. In doing so, they realigned dominant images of American engineering in a tenuous interpretation of an ideology of technological change as a philosophy of professionalism.

Contested Responsibilities

Efforts to address technology's ill effects in the member societies reveal the opportunities and limitations afforded by the ideal of the responsible professional. The rhetoric of responsibility was infinitely malleable and universally desirable, which gave reformers' legitimacy but also fostered cooptation. Reformers, moreover, struggled to define responsibilities that extended beyond client service. It was one thing to be held accountable for ethical lapses, but quite another for the social effects of technology. After accepting such obligations, what was the proper course of action? Could responsibilities be outlined for unintended effects that circumvented one's intentions? Were there metrics for predicting the unpredictable?

To capture how the shifting meanings of responsibility generated and foreclosed reform in the member societies, this chapter focuses primarily on one organization—the American Society of Mechanical Engineers. Founded in 1880, the ASME was one of the oldest and largest engineering institutions in the United States, and by 1965 it encompassed sixty thousand engineers in fields ranging from fluid mechanics to systems design. The ASME was a tax-exempt nonprofit institution governed by a voluntary national Council with an annually elected president and regional chapters. From 1955 to 1975 the majority of Council members were executive managers concentrated in the power industry. ASME engineers worked in utility, manufacturing, and aerospace industries; government research laboratories; universities; and an array of small businesses. Many of the most active members shared overlapping affiliations with sister societies such as the IEEE and field-spanning organizations such as the NSPE, the American Society for Engineering Education (ASEE), and the Engineers' Council for Professional Development (ECPD).

The ASME was not simply representative of professional responses to technology's critics; it was a pacesetter. In the late 1960s a group of reformers in the rank and file challenged the ASME to take culpability for technology's "unintended effects" and to protect its engineers in the execution of their judgment. The reformers created a Technology & Society (T&S) Committee that interrogated socio-technical problems through reflective inquiry. Their initiatives prompted the ASME Council to undertake a national program under the banner "Making Technology a True Servant of Man," directed by society president Donald E. Marlowe.

Historians of American engineering will recognize institutional ghosts in the ASME's quest for social consciousness. Beginning in 1908 and culminating in 1918, a small group of ASME reformers staged a muckraking campaign criticizing the influence of utilities companies in society governance. Led by Morris Llewellyn Cooke—a disciple of Frederick W. Taylor—reformers sought structural change aimed at professional autonomy. The committee it spawned instead produced a statement of ethical principles that "forbade nothing."[5] The ASME nonetheless became a leading promoter of the progressive ethos in the engineering profession at large. In 1930, amid celebrations of the Society's fiftieth anniversary, the engineer's moral command rolled off the pages of *Mechanical Engineering*. "The Engineer," its editor wrote, was an "apostle of progress," a "great cosmopolitan whose purview comprises the whole human race and whose ambition is to serve all."[6]

The ASME's legacy of reform fueled and restricted its second revolt, which began at the intersection of status seeking and moral philosophy. Lofty ideals of political leadership that had faded in the Great Depression were rekindled as images of responsible professionalism flourished in the post-Sputnik era. Scientific advances, occupational growth, and federal patronage generated ambitions about attracting the tens-of-thousands of mechanical engineers that were not ASME members and simultaneously provoked fears of the "vanishing" and "fading" of engineering.[7] A proliferation of articles between 1958 and 1964 with titles such as "The ASME IS a Professional Society" mixed bravado and insecurity.[8] Society officers asserted that in a technological revolution engineers should attain government advisory positions but that specialization and lack of professional unity limited their success. ASME president Ronald B. Smith, a senior vice president at M.W. Kellogg Company, emphasized the engineer's responsibility in contrast to the scientist's lack of social obligation, explaining that the engineer's duty went beyond the mere pursuit of knowledge in a method "combining disciplined judgment with intuition, courage with responsibility, and scientific competence within the practical aspects of time, of cost, and of talent."[9]

Assertions of responsibility for status enhancement were made against the background of two unprecedented outcomes of World War II that were reshaping understandings of the bonds among individuals, institutions, and technology. By upholding individual culpability for bureaucratic actions in the holocaust, the Nuremberg and Eichmann trials were formative for interpretations of professional obligation.[10] At the same time the scientists who had built the first nuclear weapons created interest groups that shaped the terms of America's nuclear debate and established models for the political mobilization of experts.[11] These "traumas of the twentieth century" contributed to reinterpretations of political philosophy and theology by Hannah Arendt, Dwight Macdonald, H. Richard Niebuhr, and others, which in turn informed civil rights activists and student organizers.[12]

When, in the mid-1960s, the ethical basis of professionalism became contested terrain, engineers engaged with these emerging interpretations of moral agency. The catalyst in the ASME was a series of *Mechanical*

Engineering articles that compared engineering with other vocations to defend its professional status. In 1964, Joseph G. Wilson, a Shell Oil manager, ranked eleven occupations according to a weighted scale of specialized training, legal responsibility, personal relationships, and ethics. Engineers lost points in specialized training, personal relationships, and ethics, but with a score of eighty out of a hundred points, lagged only doctors and lawyers in professional orientation.[13]

In response, AMSE member Victor Paschkis introduced a variation into the professionalism discourse that generated more letters to the editor than any topic in the preceding two decades. Paschkis, a Columbia University professor and Quaker peace advocate, criticized the lack of ethics in the ASME and argued that every engineer must take *personal* responsibility for his work as a necessary condition of professionalism.[14]

Challenged to turn the politics of responsibility inward, ASME engineers drew on formal and informal moral reasoning in a debate that ran from 1964 to the early 1970s. Associate member John Ryker rejected Paschkis's idea that every engineer was his own judge. "Assuming that his employers are members of legally constituted businesses," Ryker argued, "an engineer can, and in my opinion, should provide his services to anyone willing to pay his price."[15] It was especially dangerous, he continued, for an organization like the ASME to reach consensus positions.

Paschkis's supporters drew contradictory defenses from analogies to the medical and legal professions. W. E. Little, an engineer from Richmond, Virginia, for example, argued that doctors and lawyers were obligated to accept "morally repugnant" work and he thanked God that engineering's *lack* of mandated ethical codes allowed him to decline work on cobalt bombs or nerve gas.[16] Another refuted the assumption that professionals had to serve reprehensible clients, asserting that doctors in concentration camps were not obligated to carry out experiments, nor were lawyers duty-bound to assist predatory clients. The key task facing engineers was rather to identify responsibilities "corresponding to the clear obligations for doctors to save life, and for lawyers to defend accused."[17]

In reaction to this line of argument conservative and libertarian engineers asserted that it was hazardous to posit a universal morality. James

D. Thackrey of Santa Ana, California, for example, anticipated Milton Friedman's critique of social responsibility as a subversive doctrine: "I don't OWE that largest collective to which I belong, 'the public,' anything," he wrote, "and I refuse to feel guilty because some genuine or phoney collectivist thinks I should put his welfare on par with mine."[18]

Ethical codes seemingly would be the logical venue to clarify these disagreements about responsible action. From their earliest days the member societies had such guidelines, but unlike codes in law and medicine that were backed by deliberative bodies with the power to disbar members, outside of state-employed and certified professional engineers, such rules carried little punitive weight. Codes instead precluded disputes by treating corporations as individual clients and putting devotion to client above the public or individual conscience.

The renewed emphasis on professionalism sparked an ethics revival, with the ECPD releasing a major update in 1963 to its Canons of Ethics that was the outcome of a six-year conversation with the ASEE and the NSPE. Its fundamental principles, however, were little different than previous versions. "The Engineer" was to "advance the honor and dignity" of the profession through honesty and impartiality with "devotion" to "his employer, his clients, and the public" and to use "his knowledge and skill for the advancement of human welfare."[19] The new Canons grappled with corporate labor, but like earlier versions lacked consequences and reiterated loyalty to one's company over the public. The Canons also failed to mention "technology"—no minor omission in a profession that had begun to consider its livelihood at stake over the term.

Indeed, by the mid-1960s, ASME leadership increasingly understood questions of status and obligation in the context of evolving normative visions of technology. Council members referenced technology & society literature in a mixture of overt propaganda combined with the belief that engineers could be applied social philosophers. As responsibility shifted from an uncontroversial virtue for political advancement to include an interrogation of technology's social impact, ASME's leadership reframed claims about professionalism by emphasizing technology's uncertainties as proof of the need for enhanced authority. To a striking degree Progressive Era rhetoric structured their efforts. Responsibility was first and foremost a power to be discharged by the elite professional—a virtue made possible by objectivity in public affairs. Richard G. Folsom, president of RPI, for example, challenged the notion that engineers were amoral organization men:

Should engineers take the attitude that "we can get you to the moon, but don't expect us to tell you whether the trip is necessary?" Here on earth, our environment is being polluted by industrial and social waste. . . . All of these problems require superior technological information and knowledge and wisdom before a solution can be envisioned; and that solution is political, economic, and technical. If the engineer, who is the only one who has the technological information, sits idly by and does nothing, others will attempt solutions that must, perforce, be inadequate because of the lack of the basic knowledge required to get to the heart of the problems.[20]

In the face of new complexity and an awareness of unwanted results, Folsom deviated from predecessors, however, in seeing social responsibility as something foisted upon engineers against their will. No longer masters of nature and society, necessity compelled engineers to political action.

Underlying these ambitions of enlightened management was an awareness of the proliferation of competing claims over the blame and control of technology. On the one hand, *Mechanical Engineering* editorials decried the public's adulation of "the misfit, the pervert, the drug addict, the drifter, the ne'er-do-well, the maladjusted, the chronic criminal, the underachiever, the loser . . ." in contrast to "the doer" and "the succeeder" who took seriously "responsibility for achievement."[21] On the other hand, letters to the editor condemned the draft, described protestors as "patriotic young Americans full of ideals," and promoted the benefits of LSD.[22] Corporate recruiting liberally embraced social responsibility. Bethlehem Steel, for example, presented a cartoon multi-racial workforce standing beside a factory, rendered as the word "CONCERN."[23] In feature articles, executive managers likewise appealed to the rank and file as responsible professionals. David Packard, chairman of Hewlett-Packard, cited new duties to limit technology's dislocating impact but asserted that "no one will deny that profit making is, indeed, the most important responsibility of a manager."[24] In 1968 Donald Burnham, president of Westinghouse, similarly identified "productivity" as the most powerful rebuttal of technology's critics. Burnham maintained that *growth* was the engineer's "great social responsibility" and admonishments to become politicians, social scientists, or artists were empty platitudes.[25]

Graduates who are concerned about society and the quality of life ought to consider a career with Bethlehem Steel.

Consider environment. We recognize our responsibility to restore and preserve a viable environment. Few industrial corporations or municipalities can match our accomplishments in environmental quality control.

Consider equal opportunity. We are one of industry's largest employers of minority-group citizens. Our management training program includes persons of all races, male and female. What's more, every effort is made to assure that all employees have equal advancement opportunities. Where lack of education has been holding individuals back, we offer basic education courses.

We believe in people. We believe in personal development. We encourage individual participation in community and social welfare activities. We sponsor nonpartisan political education courses. We underwrite continuing education for employees, at both the undergraduate and graduate levels. We provide management training and development.

All this is more fully described in our booklet, "Bethlehem Steel's Loop Course." It's required reading for anyone who is thinking *career*. Copies are available in many placement offices; or write: Director—College Relations, Personnel and Management Development Division, Bethlehem Steel Corp., Bethlehem, PA 18016.

BETHLEHEM STEEL
An equal opportunity employer

Figure 4.1
Responsibility as a recruitment strategy
Source: Bethlehem Steel, "Concern," *Tech Engineering News* 52, no. 7 (December 1970): 20.

Debates about technology in the ASME mixed Progressive Era tropes with contemporary technology & society literature, stressing the mitigation of unintended effects as the profession's newest challenge. Shared premises about out-of-control technology nonetheless spawned divergent ideas about how to alter professional practice. Most proponents of an ideology of technological change insisted that an accusatory mode hindered solutions and that prolonged breast-beating about side effects constrained economic growth. For an activist minority, however, professionals were not excused of culpability for unintended consequences. Just because the deleterious effects of technology were not intentional did not mean they were unknowable or that they could be avoided; consequently industries that did not undertake long-range planning should bear the burden of their actions.

The extension of cost–benefit analysis to new domains provided an actionable middle ground. In a heralded article in *Science,* Chauncey Starr, a nuclear industry executive turned UCLA dean, argued that "social benefit" and "technological risk" could be balanced quantitatively. By plotting fatalities per hour of exposure to a technology against monetary benefit per person, one could calculate acceptable risk for any device or system.[26] Social risk moreover was a manipulable entity that could be modified by "benefit awareness." In other words, acceptance of nuclear power could be improved through public relations.

A steady stream of criticism in the ASME, however, introduced themes of technological politics to explain engineers' environmental abuses and lack of autonomy. In 1970 an exchange between the ASME's new president, Donald Marlowe, and its dissenting members highlights the character of unrest at its peak. In a jeremiad titled the "New Luddites," Marlowe, who was dean of engineering and architecture at the Catholic University of America, declared assaults on technology to be a crisis that challenged engineers' "very way of life."[27] While he received many appreciative responses from the rank and file, another faction accused him of mischaracterizing the nature of the crisis and urged dialogue with technology's critics. Paschkis, who shared much of Marlowe's social vision, offered ASME engineers an abridged syllabus in response. "The Luddites," he explained, "called for a clear, violent, and redundant 'no' to technology. Modern critics . . . including such men as Mumford and Galbraith in this country, Ellul in France, Juenger and Picht in Germany,

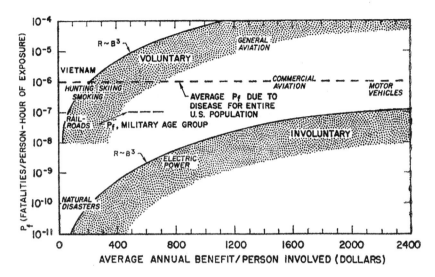

Figure 4.2
Quantification of social benefits and technological risks
Source: Chauncey Starr, "Social Benefit versus Technological Risk," *Science* 165, no. 3899 (September 19, 1969), 1234. Courtesy of the American Association for the Advancement of Science.

say 'technology as it is moving now, leads to an impossible situation; let's see how it can be directed to avoid this calamity.'"[28] Stan Weissenberger, an associate professor at the University of Santa Clara, further emphasized the limits of the engineering method as a social philosophy. "I submit, that we do have real and profound problems that technology has contributed to," Weissenberger wrote, "and which it is not at all clear that technology can completely solve."[29]

Conflict within the ASME was anathema to an organization that represented engineering's fraternal ideal. When Marlowe became ASME president, he lamented the "paranoid style" and "emotional content" of the debate.[30] Rank-and-file reformers also cautioned against partisan politics as they challenged the ASME to serve engineers in protecting the public.

Between 1968 and 1973, there were two intersecting campaigns to stabilize the ASME by organizing it toward responsible purpose. The first was led by Paschkis, who strove to transform partisanship into reflective practice from the ground up. These grassroots reformers argued that by institutionalizing critical inquiry of technology's consequences, they could influence peers and secure ethical protection from the ASME to serve the public. The second originated in the Council and was shaped largely by Marlowe. It established societywide goals to reassert engineers as technology's masters. Both targeted ASME publications to normalize their position and enroll members. Both called upon intellectual authorities to elaborate their vision of responsibility. The origins and intersections of these two efforts demonstrate how conversations about technology were channeled in engineering's largest social institutions and how the member societies were ill-equipped to act upon their ideals.

The Moral Dilemmas of Technological Change

By any account Victor Paschkis was a responsible man. Born in Vienna in 1898, Paschkis was the son of nonconformist Catholics. As a child he developed a love of politics and philosophy, reading antiwar tracts in his family library. When conscripted to the Eastern Front, he continued his pacifist studies in the trenches. His technical career began at the Technische Universität Wien, where he split time between lectures on engineering and socialist labor politics. After earning his ScD, he took a job at Allgemeine-Elektricitäts-Gessellschaft (AEG). Working in Berlin, he expanded his commitment to the peace movement in the international, interfaith group the Fellowship of Reconciliation.[31]

Paschkis first became embroiled in conflicts of moral and occupational duty as an employee at AEG. Tasked with testing transformers as they came off the line, he identified a pattern of faulty equipment. After a few days marking defective gear, his boss demanded that he falsify the results or be fired. Untold engineers have left similar meetings in a silence that can define a career. Paschkis was righteous and lucky. He successfully pleaded his case to an executive manager who was a distant in-law, and was rewarded for his honesty with a transfer and a promotion.[32] Nonetheless, the lack of formal recourse impressed upon Paschkis the need for collective representation, and he retold his story innumerable times to draw attention to the practical problems of conscience.

Paschkis encountered an even greater moral dilemma during World War II. To escape National Socialism, he immigrated to the United States and settled in New York. As a research professor at Columbia University, where he would work for almost three decades, he built the Heat and Mass Flow Analyzer, an analog computer for making thermodynamic calculations that were impossible to solve by experimentation.[33] Paschkis was a conscientious objector, who during the war volunteered as an air warden in Harlem. One day his department head asked him to run data on his analyzer, and Paschkis neglected to ask why. Only when reading an account of the Manhattan Project did he discover that he unwittingly assisted in nuclear weapon design.[34]

In the wake of World War II, Paschkis sought moral guidelines for a world of holocaust, atom bombs, and cold war fear. Scientists and engineers were not more or less virtuous than other citizens, he argued, but given technology's power the stakes of their choices were greater. Technical experts thus could not compartmentalize value judgments from their labor. Writing in the Quaker press, he claimed that every individual act had a chain of effects and no one could claim to be "only a cog in the wheel." Speaking up could be self-destructive, but acts of honest dissent would "spread the idea of personal responsibility" and generate institutional support. [35]

But an individual's moral compass could get one only so far. In 1949, at the encouragement of his friend the civil rights organizer A. J. Muste, Paschkis founded the Society for Social Responsibility in Science (SSRS). Initially intending to call themselves Scientists and Engineers for Peace, SSRS's highest purpose would be

to foster throughout the world a functioning tradition of world-wide cooperation, whereby science and technology contribute fully to the benefit of world humanity, and never to its harm or destruction; to embody in such tradition, the principle of both abstention from destructive work and devotion to constructive work; to ascertain through open and free discussion the boundary between constructive and destructive work, to serve as a guide for individual and group decisions and action. [36]

By 1960 SSRS had a few hundred members, including Max Born and Linus Pauling. Based on internationalist pacifist principles, it was a group of scientists and engineers created outside any professional body aimed at individual consciousness-raising rather than institutional reform.[37]

In the 1960s, civil rights organizing altered Paschkis's understanding of responsible professionalism in a technological age. In addition to fighting the injustices of Jim Crow, he joined an eclectic network of intellectuals, labor leaders, and scientific luminaries who linked the plight of black America to technological change. For Paschkis, the catalyst was a manifesto called the *Triple Revolution*. Produced by the Center for the Study of Democratic Institutions and sent to President Lyndon Johnson, the *Triple Revolution* represented the apex of optimism and anxiety about automation. It identified a confluence of revolutions in cybernation, weaponry, and human rights that created inequality between technology's experts and the growing mass of deskilled labor. Restoring social justice and political access, it argued, would require economic planning to manage change.[38] In 1964, at the First Annual Conference on the Cybercultural Revolution, Paschkis argued that distributing technology's benefits required a moral revolution. Drawing examples from his own automation research, he argued that minorities were the worst casualties of technological change because discrimination limited their ability to retool for new jobs. To resolve the paradoxes of innovation, he argued that affirmative action programs and job training should be developed alongside technical improvements.[39]

Paschkis became an avid interpreter of technology & society literature, integrating its lessons into his moral vision. As emeritus professor at Columbia, he arranged a seminar series for faculty and students on "ethics in an age of technological change," inviting speakers from medicine, law, and engineering.[40] He also lectured on responsibility at the nation's engineering schools—including Berkeley, Illinois Institute of Technology (IIT), MIT, Purdue, and Wisconsin—and internationally in Austria, Belgium, Czechoslovakia, East and West Germany, Holland, Norway, Poland, Turkey, and Yugoslavia. In 1970, he and nine other Columbia faculty members proposed a technology & society course that put Mumford and Ellul in conversation with Harvard Program projects and speeches by prominent technologists.[41]

Paschkis came to believe that engineering reform required a wholesale rethinking of the nature of technology. His approach took shape as an amalgamation of an ideology of technological change and deontological moral philosophy. Like many reformers he tried unsuccessfully to publish a technology & society monograph. His manuscript criticized hippies and the New Left for lacking viable alternatives, but equally chastised engineers and social scientists who blithely quantified human values, concluding that the challenge of modernity was "to have the megamachine but not become subservient to it." He recognized unintended

TO FOSTER BETTER UNDERSTANDING AND CONCERN
ABOUT THE IMPACT OF TECHNOLOGY ON OUR SOCIETY,
THE DEPARTMENT OF ENGINEERING AT SAN FRANCISCO
STATE COLLEGE TOGETHER WITH THE DEPARTMENT OF
INTERDISCIPLINARY SCIENCES, AND THE SOCIETY FOR
SOCIAL RESPONSIBILITY IN SCIENCE, PRESENTS

Victor Paschkis

WHO WILL COMMENT ON HIS RECENT EXTENSIVE LECTURE
TOUR IN COUNTRIES OF EUROPE AND THE NEAR EAST,
INCLUDING

ISRAEL	YUGOSLAVIA	AUSTRIA
POLAND	CZECHOSLOVAKIA	EAST GERMANY
WEST BERLIN	NORWAY	ENGLAND
BELGIUM	HOLLAND	WEST GERMANY
	TURKEY	

DR.PASCHKIS WILL REVIEW THE REACTIONS OF ENGINEERS
AND OTHER CITIZENS TO THE CHALLENGING POINTS HE
RAISED IN THE COUNTRIES HE VISITED. HE WILL
RELATE THOSE REACTIONS TO THE PROBLEMS NOW CONFRONT-
ING US IN THE UNITED STATES.

SCIENCE 101, SAN FRANCISCO STATE COLLEGE

THURSDAY, NOVEMBER 13, 1969   7:30 P.M.

ADMISSION IS FREE. COMPLIMENTARY REFRESHMENTS WILL
BE SERVED BY THE ENGINEERING SOCIETY, A STUDENT
ORGANIZATION.

Figure 4.3
Moral responsibility for out-of-control technology
Source: Victor Paschkis, "Society in Revolt," box 6, Victor Paschkis Papers,
Swarthmore College Peace Collection.

consequences as the preeminent problem of the technological age, but sought to identify moral duties for their evaluation and equitable structures for deciding who would bear what burdens. Expertise was essential for resolving socio-technical issues, but only with the representation of affected parties and punitive outcomes for those who neglected unintended effects in their work.[42]

The Division of Technology & Society

If engineers were to respond as moral professionals to socio-technical crises that were only vaguely understood, they needed to define the relevant problems, generate conceptual tools, and justify their authority to multiple constituencies. In 1968, Paschkis was invited by the Aviation and Space Division to arrange technology & society sessions at the ASME's annual meeting. He used the event to form a permanent Technology and Society Committee that channeled dissenters as well as ASME Council members.

The T&S Committee attracted some five hundred engineers representing every regional section of the ASME. Its membership was more academic than the ASME at large, with a quarter of its participants working or studying in engineering schools.[43] Religious faith also was high. Hendrik B. Koning, a Philadelphia Electric Company employee, for example, was an ordained Episcopal priest performing charitable work with high school students. Motives for joining ranged from evaluating "what technology and engineers might do to preserve a humane society" to "shield[ing] the ASME and all Engineering Societies from the wild claim that our environmental and social problems are the result of our rapidly developed technology" by "constantly remind[ing] the public that our technology was invented and developed at a rate greater than society could assimilate."[44] The T&S Committee's inaugural event sessions likewise presented a spectrum of belief about responsibility. Kurt H. Hohenemser, a member of Barry Commoner's Committee for Environmental Information, argued that scientists and engineers should participate in controversies by presenting information without bias, but they should not make value judgments. At the same time Marlowe cited Starr's work on risk assessment to contend that disputes of profit and public interest could be managed by engineering methods.[45]

With Paschkis at the helm, the T&S Committee framed unrest in terms of professionalism instead of activism and dedicated itself primarily to labor conflicts arising from technology's unintended effects. Paschkis argued that "responsible technology" was possible if engineers acted with conscience, but until the idea of unintended effects received broad acceptance, management could dismiss concerned engineers. He outlined how member societies could implement conflict resolution to protect engineers in confronting the technology & society interface, and asserted that if the ASME did not take such steps, it should eliminate any mention of a "profession" in its publications.[46]

The T&S Committee advocated technology assessment and warned that America's unchecked growth was unsustainable. Paschkis and other committee members believed that dramatic lifestyle changes would be needed to address environmental degradation and poverty. Assessment was the first step for recognizing the requisite tradeoffs. In a response to the NAE's congressional assessment report, Paschkis reworked excerpts from his technology & society manuscript into a manifesto titled "Assessment—By Whom, for Whom?" Against the charge that critical reflection slowed innovation, he responded with incredulity that technology was on a path to "chaos":

We encourage developing countries to catch up with us and simultaneously increase our own consumption. The increasing avalanche of gadgets not only increases the consumption of irreplaceable natural resources, but also presents the psychological hazard of being "habit forming." Thus, if slowing down of "progress" should come as a byproduct of a necessary new step of assessment, welcome it![47]

Ambivalent about technocratic solutions, he nevertheless concluded that the only way to account for nonmonetary factors in design was to "evaluate the valueless" as objectively as possible, because social ethics had deteriorated to such a level that they would have little impact.

The T&S Committee, in short, took technology's critics seriously and intended to make an ideology of technological change a means for progressive control. Its members achieved an important milestone in 1972 when they petitioned to become a technical division with equal standing to much larger branches such as Production Engineering. The new T&S Division also was granted a monthly column in *Mechanical Engineering* titled "The Socio-Technical View." With its newfound institutional status, the T&S Division drafted a Canon of Ethics that emphasized the "honor and dignity" of engineering, identified obligations for "undesirable consequences," and upheld the right to individual conscience.[48] The Division's approach to out-of-control technology, however, was overshadowed by a parallel effort to create responsible professionals.

The ASME Goals

Transforming the ASME from the grassroots was perhaps less complicated than altering it from above. The sincere reformer on the Council knew better than most the limitations of what the member societies could achieve. Participation in the ASME was voluntary and annual dues provided the rank and filers with a sense of ownership over Council decisions. The budget those dues generated was just enough to host annual meetings and publish technical proceedings. A presidential term lasted a single year. ASME members came from radically different backgrounds and held antithetical positions about what engineering was for. The corporate institutions that stood to be reformed were also the greatest constituency and source of authority in the member societies. Indeed one's social circle included those companies' top managers. Nonetheless, to ascend to a position of leadership over sixty thousand technical professionals came with optimism about the power to change the world.

To assume the national offensive and repudiate the new Luddites required the intellectual's penetrating inquiry, but it also demanded faith in the engineer's ability to solve any problem. At the same time that Paschkis was building his coalition, the Council addressed the problems that beset it by rallying around a set of "great centralizing ideas" known as the ASME Goals. The Overriding Goal displayed a commitment to tame out-of-control technology:

To move vigorously from what is now essentially a technical society to a truly professional society sensitive to the engineer's responsibility to the public, and dedicated to a leadership role in making technology a true servant of man.[49]

In contrast to the ASME's existing ethical codes, the Goals put the rhetoric of public obligation before those of corporate and member interests. While the Goals echoed well-tread formulations of professionalism, its prose borrowed heavily from Paschkis' T&S Committee, making reference to "deleterious 'second order' effects of technology," "mass layoffs," and "technological imperialism."[50]

ASME Goals: Making Technology a True Servant of Man

Overriding Goal: To move vigorously from what is now a society with essentially technical concerns to a society that, while serving the technical interests of its members ever better, is increasingly professional in its outlook, sensitive to the engineer's responsibility to the public's interest, and dedicated to a leadership role in making technology a true servant of man.

Goal 1. To broaden the horizons of individual engineers and foster effective means by which they may relate their work and talents to the interests of society, and to provide support for them in the exercise of their societal responsibilities.

Goal 2. To develop procedures for establishing an ASME position on public issues to which engineering views are relevant, and for the expression of such views by duly authorized spokesmen.

Goal 3. To foster communication among engineers, the other professions, and the public for mutual understanding of the true role and contributions of technology.

Goal 4. To develop in ASME a responsiveness to and concern for the economic and professional needs of its members, in addition to its customary concern for their technical needs.

Goal 5. To develop a closer and more effective relationship among engineering practitioners, those who educate for the profession, and those who are being educated.

Goal 6. To take a leadership role in instilling individual recognition of the obligation to maintain and expand competence and in encouraging the various technical disciplines within mechanical engineering to take responsibility for the nature and quality of continuing education in its area.

Goal 7. To encourage membership and participation in ASME of all those engaged in mechanical engineering.

Goal 8. To extend the scope and to increase the timeliness, pertinence, and effectiveness of all ASME communications media to engage the interest of, and fully inform, all elements of the Society.

Goal 9. To undertake, on a continuing basis, an assessment of the priorities for research, and identification of new or unfulfilled needs.

Goal 10. To play an even more active and effective role in the generation and continuous improvement of codes and standards under the Society's jurisdiction or in cooperation with other agencies (both domestic and international), and to increase our efforts toward implementation of such codes.

Goal 11. To provide government at all levels with advice on engineering matters and policies affecting the public interest, and to develop a climate of understanding and credibility that will foster a continuing dialogue.

Goal 12. To direct ASME efforts in all possible ways toward a unification of the engineering profession, increasing its strength throughout the entire continuum from the campus onward, thus making it a more effective force in relations with the public, the government, and industry.

Goal 13. To strengthen through reciprocal cooperation with indigenous groups the ties between engineers in the U.S. and those in all other countries, for wider dissemination of engineering skills and more effective interchange of technology.

Goal 14. To encourage and assist the entry into the engineering profession of all capable individuals who because of race, sex, creed, or economic handicap might otherwise find it difficult.

Figure 4.4
ASME Goals
Source: "Goals: Basis for Action Programs, 'Making Technology a True Servant of Man,'" *Mechanical Engineering* 93, no. 4 (April 1971): 16–19. Courtesy of ASME.

First suggested by outgoing ASME president George F. Habach, the Goals were implemented under Marlowe's leadership. Marlowe had been an advocate of invigorated professionalism throughout his career. He was active in the NSPE, and in the early 1960s wanted to make the rank of "full member" in the ASME contingent on professional registration. At the same time he recognized that most engineers did not work in client relationships and lacked fiscal authority. He worried that federal attempts to mitigate technology's negative effects would result in "socialized engineering," arguing that the profession "must recognize, as it never has before, its professional responsibility to become the principal advisor to the government" in order to set regulations and standards.[51] Doing so required convincing engineers to choose public service rather than private industry.

Unlike the majority of presidents who preceded him, Marlowe was an academic rather than an industrial manager. A prolific writer, he cited a litany of intellectuals to contextualize the engineer in an era of accelerating technology. His inspirations ranged from David Ricardo and Mary Shelley to contemporary historians such as Melvin Kranzberg and Edwin T. Layton Jr. As useful as these texts were for defining the engineer's social vision, he stressed that only quantitative methods would resolve socio-technical problems.

Marlowe argued that the nation's turmoil resulted from the fact that it was feasible to resolve modernity's most vexing problems, but that no one yet knew how to mobilize new technology for social needs. He considered himself a Jeffersonian democrat and was intent on making the ASME a force of social change. In his 1969 essay "Prometheus Unbound," he asserted that as a "by-product" of their history engineers had come to possess "an extraordinarily powerful methodology for the solving of problems." Unlike the mere "intuition" of the average citizen, the predictive, and thus superior, "engineering method demand[ed] years of preparation and effort."[52] In another *Mechanical Engineering* feature article he argued that socio-technical problems could not be left to the "soft sciences" because the mathematics was too complex. Engineers could forge a path to leadership by opposing the "intuitionists," developing systems theory, working with social scientists, rekindling "historic service" to government, resisting the "temptation to power proffered by the fascist ideology," reforming engineering education, garnering managerial independence, and attaining policy-making positions.[53]

Marlowe put his stamp on the Goals, but they were a collective project. A committee of ninety engineers drafted the preliminary version by applying systems techniques developed for consensus building in urban renewal projects.[54] The group included six past and eight future ASME presidents as well as the editors of *Product Engineering*, *Power*, *Electrical World*, *Chemical Engineering*, and *American Machinist*. So that the Goals would be the result of the ASME as a whole, rank-and-file members were represented at all planning stages. In a yearlong process the Goals were presented for revision in over one hundred meetings, and after four thousand votes were tallied, the majority of Goals passed with close to 90 percent approval.[55]

The Goals Statement was an aspirational document with the weight of the ASME Council and its president. The Council created working parties intended to provide dialogue with the public, design portable pensions, and draft guidelines for supporting individual engineers in the execution of their social responsibilities. But the Goals gave the most tangible support to enhancing the profession's federal profile. Emphasizing long-standing excellence in boiler standards, Council members argued that the ASME could extend its purview to environmental codes.[56] Marlowe collaborated with Congressman Daddario to draft a handbook that introduced engineers to legislative procedures, and *Mechanical Engineering* published a monthly column on government trends. Amid talk of abandoning its tax-exempt status, the ASME established a Washington office and initiated a Congressional fellows program subsequently copied by other engineering member societies.

Normalcy and the New Professionalism

In 1972 the ASME appeared to be fulfilling its historic mission. Recently elected society president Kenneth A. Roe declared in his annual report that "professionalism is, of course, not new. What is new . . . is the transformation that is being wrought in the goals that are being striven for and by the responsibilities that are being thrust upon the professional." Nodding to the T&S Division, he outlined a vision of responsibility for technology's unintended effects. Because of the relative youth of the profession, engineers should not be blamed for the "cumulative ill effects of [their] ingenuity."[57] At the same time they stood best suited to control technology in the future.

But the tangible impact of the Goals and the T&S Division was not what its advocates had hoped. The working parties assigned to study social responsibilities did not result in ethical protection in traditional conflicts of interest, much less in new socio-technical realms.[58] Moreover, despite its effort to engage all ASME engineers, less than a tenth voted on the Goals.

While the Goals failed to expand the dominion of mechanical engineers, they succeeded in dampening the ASME's sense of existential crisis. To generate credibility among future professionals, the ASME held essay contests in which student members were asked, "Do you think the engineer should assume social responsibility?" Winning papers mimicked the language of the Goals.[59]

The normative function of the Goals is best seen in the ASME's position on alternative fuels. In 1971, as the Goals were solidified, debate shifted from philosophizing about technology writ large to the energy crisis. Here was a socio-technical controversy in which mechanical engineers felt confident about the line between ideologue and expert servant. Calls for resource planning and conservation abounded, as meetings investigated solar, geothermal, and wind power. With thousands of its members employed in the energy sector, however, the ASME overwhelmingly supported nuclear power and coal gasification. In December 1975 it issued its first official policy statement mandated by the Goals to champion oil, natural gas, coal, and nuclear as the "only significant energy sources available" for the rest of the century.[60]

The T&S Division reformers at first saw the Goals as a boon to their cause. Most agreed with the aspirations of the Goals and wanted to make sure they did not become platitudes. They offered extensive revisions to the preliminary draft, recommending that its general "broadening of horizons" stress concrete responsibilities for unintended effects and establish conflict resolution provisions to aid engineers who took such considerations against employers' wishes.[61] Though the T&S Division's suggestions were rejected, its members referenced the Goals statement for years after it dropped out of conversation elsewhere in the ASME.

Despite its symbiotic relationship with the Goals, the T&S Division held a tenuous position. Its blurring of boundaries between the social and technical created enemies in the ASME, and, in 1974, a Policy Board review attempted to break the T&S Division into two "nontechnical" committees.[62] Members defended their status as a technical division by stressing technology assessment and the ASME's public image.[63] "The Socio-Technical View" column brought the T&S Division's voice to sixty thousand readers, but the Division was unable to expand its membership beyond a thousandth of ASME engineers. Undeterred, Paschkis contributed to a resurgence of debate about professional ethics in the late 1970s and attempted unsuccessfully to persuade Theodore Roszak to speak at the ASME's 100th anniversary in 1980.[64] The T&S Division exists to this day, with 6,000 members, but without the moral exigency present at its formation.

Conclusion

The lens of professionalism continues to shape prevailing understandings of American engineering among both engineers and those that study them. There is ample evidence, however, that the professional ideal misrepresents American engineers in the aggregate. This observation was not lost in the member societies. Indeed, in his capacity as consultant to the ASME Goals, William H. Wisely, the executive secretary of the ASCE, questioned the very notion of an "engineering profession," declaring it to be a "heterogeneous nebula of vague peripheral definition" within which there was a "helter-skelter mosaic of hundreds of specialists" connected by "superficial" bonds.[65]

For reformers in the late 1960s and early 1970s, responsible professionalism nonetheless seemed the most effective path toward the control of technology and renewal of self. Because of Paschkis's contribution to a grass-roots redefinition of responsibility and Marlowe's embrace of the socio-technologist interpretation of an ideology of technological change, the ASME's new professionalism went further than any other member society in the United States. But the ASME was not unique. Paschkis helped his Columbia colleagues Stephen Unger and Mario Salvadori undertake similar movements in the IEEE and ASCE. The Goals Program likewise was carried out in consultation with the ASCE, NSPE, and others.

Across every member society and in the public relations of the corporations that employed their members, engineers were urged to assume responsibility for the socio-technical world. Admonitions came both from the rank-and-file and society officers. In the AIChE, for example, reformers in its South Texas Section established a "Pollution Solution Group," while national leadership asked a new Environmental Division to investigate ethical codes for unintended environmental consequences. The report pushed for a review board to protect engineers, but the Council abandoned the effort and subsequently resisted similar measures.[66] Likewise a 1972 IEEE survey of forty-seven thousand members was two-to-one in favor of "becoming more active in political and economic matters." The survey, however, did not address moral responsibility, and was oriented toward lobbying, career planning, and public relations.[67]

The great majority of these efforts were more heat than light. Issues of unemployment found the most traction, with calls for congressional lobbying to provide assistance for retraining in environmental engineering. But even these efforts amounted to a reduction in dues for members down on their luck.

Member society reformers achieved limited success for a combination of reasons. First, the societies were ill-equipped to implement a plan like Paschkis's or Marlowe's. Their emphasis on professionalism derived not from their function as keepers of standards or as defenders of the public but rather from their role as arbiters of vision. They organized meetings, compiled and published journals and other technical literature, promoted specific forms of education, and generally worked to define a social identity. The Goals reconstituted an ambiguous image of engineering as progressive service that diffused critical inquiry. The ASME's Council and publication editors were open to healthy disagreement, but within limits. Rank-and-file reformers employed examples of atom bombs and Nazi gas chambers, but most official publications avoided discussion of Vietnam.

Second, the ethos of responsible professionalism was encumbered with a historical legacy that verged on nostalgia. Disillusioned engineers lamented this backward looking impulse. In response to the ASME Goals, for example, an angry mechanical engineer attacked its assumptions: "You are just not with it, goals committee, and there must be a whole generation or two of young engineers that are laughing their sides (or other parts of their anatomy) off at your inability to come to grips with the real problems of the day in the plain rhetoric of the day."[68]

Likewise civil engineers criticized what they saw as the ASCE's half-hearted reconciliation with the crisis of technology. "With less emphasis on praising half-forgotten covered bridges in Ohio or Iowa as national civil engineering monuments and more emphasis on admitting that New York City slums, Los Angeles smog, and Appalachian strip mines are national civil engineering disgraces," wrote one, the ASCE "would serve society as a whole rather than the narrow interests of the profession."[69]

Third, the language of responsibility at the core of reformist visions was especially prone to cooptation. In his appeal to professionalism, Paschkis depoliticized the ASME's debate about out-of-control technology. This was a point of convergence with corporate interests that sought to weaken lines of responsibility by stressing the inevitability of unpredictable side effects as a natural condition of technological change. From their perspective, excessive inquiry into the nature of such effects would hinder engineers' responsibility to the productivity of the system they served. Attempting to foresee every negative effect of technology, they contended, was a task of impossible hubris.

Fourth, an ideology of technological change was a managerial philosophy, but not necessarily a professional one. Marlowe and Paschkis interpreted unintended effects as an occupational duty and tried to enroll others in that vision. Marlowe did so by doubling down on the dominant images of engineering's past: "Much of our recent social turmoil," he argued, "arises from man's realization that he can, in fact, remake the world."[70] Paschkis, on the contrary, identified unintended effects as an unavoidable problem in *all* engineering work that inevitably created moral dilemmas.[71]

Finally, if visions of accelerating change fostered ambitions of professional leadership, they also could destabilize conceptions of mastery. To suggest that technology had negative, unintended consequences was to recognize its autonomous agency and the engineer's lack of control. A 1972 editorial in *IEEE Spectrum* highlighted the unease by citing Toffler's concept of future shock. "The challenge of the decade for electrical engineers will be adaptability," Donald Christiansen wrote. "Perhaps it is a challenge some of us cannot meet. . . . It may not be within the scope of the engineering mentality to do so."[72] From the corporate manager's perspective, it was the engineer's obligation to accept the unease as

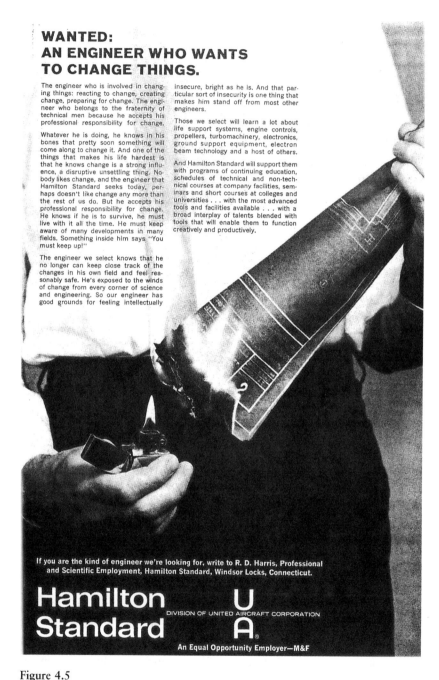

**WANTED:
AN ENGINEER WHO WANTS
TO CHANGE THINGS.**

The engineer who is involved in changing things: reacting to change, creating change, preparing for change. The engineer who belongs to the fraternity of technical men because he accepts his professional responsibility for change.

Whatever he is doing, he knows in his bones that pretty soon something will come along to change it. And one of the things that makes his life hardest is that he knows change is a strong influence, a disruptive unsettling thing. Nobody likes change, and the engineer that Hamilton Standard seeks today, perhaps doesn't like change any more than the rest of us do. But he accepts his professional responsibility for change. He knows if he is to survive, he must live with it all the time. He must keep aware of many developments in many fields. Something inside him says "You must keep up!"

The engineer we select knows that he no longer can keep close track of the changes in his own field and feel reasonably safe. He's exposed to the winds of change from every corner of science and engineering. So our engineer has good grounds for feeling intellectually insecure, bright as he is. And that particular sort of insecurity is one thing that makes him stand off from most other engineers.

Those we select will learn a lot about life support systems, engine controls, propellers, turbomachinery, electronics, ground support equipment, electron beam technology and a host of others.

And Hamilton Standard will support them with programs of continuing education, schedules of technical and non-technical courses at company facilities, seminars and short courses at colleges and universities . . . with the most advanced tools and facilities available . . . with a broad interplay of talents blended with tools that will enable them to function creatively and productively.

If you are the kind of engineer we're looking for, write to R. D. Harris, Professional and Scientific Employment, Hamilton Standard, Windsor Locks, Connecticut.

**Hamilton
Standard**   **U A**

DIVISION OF UNITED AIRCRAFT CORPORATION

An Equal Opportunity Employer—M&F

Figure 4.5
"Wanted: An Engineer Who Wants to Change Things"
Source: Hamilton Standard, "Wanted: An Engineer Who Wants to Change Things," *Mechanical Engineering* 88, no. 6 (June 1966): 145.

natural. In a particularly stark example of the responsible employee strand of an ideology of technological change, a recruitment advertisement for the aerospace firm Hamilton Standard explained how it was a moral duty for engineers to suppress critical reflection. Against the background of an anonymous, faceless engineer setting fire to his blueprints, it defined the new professionalism:

> The engineer who is involved in changing things: reacting to change, creating change, preparing for change. The engineer who belongs to the fraternity of technical men because *he accepts his professional responsibility for change. . . .*
>
> And one of the things that makes his life hardest is that he knows change is a strong influence, a disruptive unsettling thing. Nobody likes change, and the engineer that Hamilton Standard seeks today, perhaps doesn't like change any more that the rest of us do. But *he accepts his professional responsibility for change.* He knows if he is to survive, he must live with it all the time. He must keep aware of many developments in many fields. Something inside him says, "You must keep up!"[73]

The gesture of the cupped lighter in hand hinted at a desire for rebellion, but the flames were lit under duress. "Changing things" was not motivated by social progress but rather to keeping pace. Indeed one's lifeblood depended on it. Instead of responsible citizens, engineers were recast as surrogate agents without the capacity to judge.

Writing as the ASME reformation unfolded, Layton saw "sterile status seeking" and "prestige politics" in the same engineers that had given Thorstein Veblen visions of revolution. The notion of engineers as revolutionaries was doomed from the start, Layton concluded, because radical politics would "violate the elite premises of professionalism" and because engineers would still be bureaucrats. But Layton was an optimist. He concluded that the collapse of progressive ideology in the 1960s might generate "social awareness" among engineers.[74] Writing a decade later, the Marxist historian David Noble proffered no such sentimentality, but he too hinted at the potential for opposition among American engineers, which he believed would come from outside the professional mainstream. "Tied as they were to the industrial organizations which controlled the means of professional practice," Noble wrote, reformers "had to choose between being radical and being engineers."[75]

5

The System and Its Discontents

America has fashions in dirty words (or phrases) and good words (or phrases). Examples from the not very distant past are "motherhood" (good) and "red" or "commie" (dirty). It would seem that recently "ecology" has been substituted for "motherhood." . . . "Peace" is, of course, a good word—as it has always been— but it is mouthed with as much understanding and purpose as were the words "Liberty, equality and fraternity" by the mobs of the French Revolution, the very mobs which were engaged in the destruction of all three. "Military industrial complex" is, I would estimate, our favorite dirty phrase of the present time.
—John Dubbury, IEEE Transactions on Aerospace and Electronic Systems[1]

Service is one of the few universal values of engineering. The notion of service gives meaning to the contractual bonds that define what engineers are for. It connects the individual to a cause greater than self, positing that because the engineer possesses special knowledge he must use it for the greater good. The ubiquity of a service mentality is also a consequence of modern technology; the teamwork and specialization necessary to produce and maintain large-scale technological systems is facilitated by tacit and explicit dedication to a common vision.[2]

In cold war America, engineers proclaimed their service to a number of overlapping virtues and necessary evils. In corporate retreats and company songs they extolled the glory of IBM or GE. When they read about obstructionists and dropouts, they defended the system that had built an affluent society. As many as 70 percent of all American engineers directly or indirectly served the production of weapons research, development, and manufacturing.[3] In industry, academe, and government these engineers were protecting democracy against its communist enemies. They were assuring financial security for themselves and their families. By pursuing exciting technical problems they served their curiosity and the advancement of technology. If they did not perform these tasks, someone else would.

Throughout the 1960s, most engineering reformers interpreted service ideals through the lens of professional responsibility. They saw a country plagued by socio-technical problems and an engineering profession failing to meet them. To combat the inevitable "side effects" of techno-logical change, they avoided one of the main themes of technology's critics—the destructive and increasingly autonomous power of advanced weapons systems. Even Victor Paschkis, a committed pacifist, did not raise engineering's military orientation as a primary concern of the ASME's Technology and Society Division. In the few notable cases in which controversy over the defense economy erupted, it was met with resistance and epistemic closure.

Devotion seen from another angle is servitude. The testimonial of former missile engineer Bob Aldridge, published in the underground journal *Spark*, is a helpful starting point for identifying how disgruntled engineers came to view their profession's collaboration in the cold war order as a moral and structural emergency. Aldridge began his call to resistance, "The Forging of an Engineer's Conscience," by recounting a debate with his daughter—"an amiable confrontation which went until 5:30 the next morning"—in which he contended that the Poseidon mis-siles he helped design were maintaining the peace. Seeded with doubt, however, his position on the "inside" at Lockheed led to a complete reversal in which he identified the arm's race as a "struggle for bureau-cracy and profit." Engineers were "inventoried and manipulated like a machine," he wrote, "conditioned for corporate life by [the] educational system." So-called defensive weapons moreover had increased to "mon-strous" proportions that made nuclear war all but inevitable. Distressed, Aldridge joined the peace movement and began reading about "multina-tional corporations and the Third World." Initially, he confronted the nation's technological priorities from the security of his career. He hoped to raise questions in the company newspaper, and even argued his case with Lockheed Missile and Space Company president Stanley Burriss. He formed a study group on converting aerospace technology to "benefi-cial uses," but when Lockheed received more defense contracts, col-leagues lost interest. He concluded that the driving force of engineers' "self-delusion" was fear of losing economic security in a recession that itself was a product of the system. Ultimately, he decided that it was impossible to build "bombs for a living" while campaigning for "peace as a hobby." Still he maintained faith in his vocation, imploring those on the fence that working engineers could form a "powerful lobby."[4]

Few engineers were as reflective as Aldridge and fewer still took the steps he did, but his path to a career outside the professional mainstream was no anomaly. Outside and on the margins of the member societies, reformers and self-styled radicals create alternative ideals of engineering service. In describing his journey from "complicity" to a "new life," Aldridge wrote as a missionary, connecting his own experience to a general critique of technological production with the intention of raising the consciousness of his peers. His story captured the essence of opposi-tional critique—a demonstration that the way things are is not just illusory, it is a cage. This transformation involves the initiate's own expe-rience of repression combined with the appropriation of concepts that describe the processes, boundaries, and failures of the existing system. Aldridge attributed his own awakening to a confluence of familial, occu-pational, and religious factors rather than to any formal ideology. None-theless, his declaration of new purpose—"I am a dropout from the military-industrial complex"—offered an instantly understood hook for engineers in search of a different future.[5]

The military-industrial complex was the "dirty phrase" that linked cold war engineering and scientific labor with critical theories of technol-ogy. It described an apparatus of defense contractors, government agen-cies, and academic research policies but also was a shorthand for antidemocratic technological momentum in which engineers and scien-tists were complicit. In his presidential farewell address, Dwight D. Eisenhower famously named the system, lamenting that an "immense military establishment" had become linked with the nation's largest companies, which he attributed to a "technological revolution" neither of his making nor the product of a new class of scientist-politicians.[6] By labeling the complex, Eisenhower provided a legitimating vocabulary for antiwar critics. In his *Pentagon Capitalism,* the economist Seymour Melman—employed as a professor of industrial engineering—documented how interlocking institutional practices had transformed the government into an industrial manager of violence. Because of the decisions of policy elites, the United States had fallen behind other nations in the design of high-speed trains, health care, and machine tools, which had diminished "productive" opportunities for scientists and engineers.[7]

For theorists of technological politics the defense industry was the most visible example of the centralizing power of technology. It required massive commitments of financial and human capital and shaped patterns of everyday life. To Lewis Mumford, the complex represented an "ideal broth" of "megatechnics" whose maintenance required "a permanent state of war."[8] SDS activists Marc Pilisuk and Tom Hayden insisted that: "American society *is* a military-industrial complex" with a destructive character "inextricably imbedded into the mainstream of American institutions and mores."[9] Overturning the system necessitated individual and collective liberation that Herbert Marcuse argued was possible only after gaining "consciousness of servitude" and forgoing the "false needs" of status and security.[10]

This chapter explores the visions of technology and self among engineers who labored to direct their profession away from defense applications. I ask the same question that occupied trade journals, academic observers, and dissidents: How far were engineers willing to take beliefs that involved collective opposition and alternatives to responsible professionalism? Most reformers who deployed critiques of the military-industrial complex claimed that technology could serve society by redirecting federally sponsored research to domestic ends. These engineers dedicated careers to altering institutions and developing fields such as environmental engineering but did not necessarily question state-capitalist innovation or reject professional values. A tiny fraction, however, posited that once technical workers recognized their servitude, they needed to resist; to do anything less was to collaborate in the system's destructive imperatives.

Being Radical and Being Engineers

In 1972 readers of *Machine Design* learned about an alphabet soup of scientific and technical advocacy groups from the "relatively conservative to the lunatic fringe." These included networks of unemployed engineers in the Association of Technical Professionals and the Self Help Action Group, to "leftist" dissidents such as Science for the People. Describing their picketing, speech-making, education programs, and corporate boycotts, the author of "The Activist Engineer: Look Who's Getting Involved!" nonetheless doubted the seriousness and demographic reach of engineering activism, arguing that politics could be a "potentially dangerous indulgence."[11]

In its dismissive curiosity *Machine Design* put itself in a long tradition identifying engineers' potential for—but absence of—political awakening. Four decades earlier Thorstein Veblen established the prevailing explanation, arguing that engineers were a "harmless and docile sort, well fed on the whole, and somewhat placidly content with the 'full dinner-pail' which the lieutenants of the Vested Interests habitually allow them."[12] Sociologists in the 1960s and 1970s added character traits of careerism, adherence to ideals of objectivity, and social conservatism.[13]

In addition to rehashing the professional barriers to change, *Machine Design's* exposé touched on four major characteristics of engineering dissent. First, it was at the grassroots of the nation's campuses and corporate research laboratories, rather than the member societies, where criticisms of the technological order originated. Universities were leading partners in the military-industrial nexus, providing knowledge and training personnel for the cold war state. But they were also key sites of oppositional politics. Academic engineers not only were in closer proximity to jeremiads against "the system," in some cases they could walk to a lecture hall or have coffee with technology's leading critics. To a lesser extent a similar intellectual and political openness existed among engineers in companies along Massachusetts' Route 128, in greater-metropolitan New York, and in the San Francisco Bay area—regions that combined an educated workforce, liberal communities, and an economic recession that heightened engineers' sense of a failed system.

Second, ideals of service had become a fault line of dissent among technical practitioners. Reformers and radicals wrestled with identifications as *scientist* and *engineer*, *employee* and *professional*, *expert* and *activist* that shaped their rhetoric, organization, and goals. Despite the fact that professional societies appeared incapable of reforming engineering, they were not easily abandoned. The attachment was in part strategic. For activist engineers, member societies were the largest source of recruits, and their publications could reach an audience of tens of thousands. But the pull of the member societies was also about attachment to the ideal of the responsible professional. Additionally boundary disputes between scientists and engineers extended to accusations of blame for out-of-control technology.[14] Dissenting scientists and engineers were close collaborators, but posited different critiques of the military-industrial complex. Broadly speaking, scientists' posited a return to a lost state or characterized technoscience as an emergent phenomenon that made the nonneutrality of science obvious and demanded the scientist's intervention. The engineer's opposition, alternatively, began from a falsification of a myth of heroic mastery and a critique of corporate service, which required protections for dissent and a redirection of labor toward humane design.

Third, engineering activism had a distinct affective and ideological cast, which *Machine Design* characterized as "different shades of 'antiestablisment-ism.'"[15] Engineers who attempted to redefine their service did so by contextualizing the complex with a vocabulary that interpreted professional identity, labor conditions, and moral responsibility in a new light. Aldridge's account, for example, appeared in a forum with articles that referenced Mumford's *Pentagon of Power* and Marcuse's *One Dimensional Man* and critiqued "anti-human science and technology" and the "pet panacea . . . [of] professional responsibility."[16]

Finally, the local institutional and intellectual factors underlying engineers' activism ranged greatly, amounting not to a unified movement, but rather a patchwork of efforts to restore progressive meaning to engineering. Differences in sophistication and political allegiance ranged widely. Moreover, to actually change one's practice presented a multitude of problems. The system they challenged relied on specialization, compartmentalization, and secrecy. Convincing one's peers usually was a losing battle that contributed to feelings of isolation. To opt out was potentially self-destructive financially, socially, and psychologically. However, in certain situations to appropriate a critique of the system offered concrete opportunities: changing one's research could mean continued federal support in an economic recession, investigating technology's social implications could provide institutional roles for marginalized engineers, and confrontational politics could bring temporary celebrity and moral superiority.

Engineers' efforts to redefine their bonds of service, in short, were complex and heterogeneous, with dramatically different personal and professional consequences. The nexus of ideology, identity, and local circumstance is thus best understood through a series of episodes documenting the spectrum of praxis from centrist conversion to revolutionary rhetoric. In the accounts that follow, I chart the challenges of overt politics in engineering first in a small research laboratory at MIT that became a national model for "conversion," then at Princeton University where an engineering professor battled against sponsored research while transitioning toward a socio-technical career, and finally in the alternative press of the Committee for Social Responsibility in Engineering in which Aldridge found his voice.

Conversion and Opportunity

Few events encapsulated the postwar shift in perception of science and technology as completely as the 1969 work stoppage at MIT and the controversy over its special laboratories. In 1969 the Institute had a budget of $217 million of which 80 percent went to goal-oriented, sponsored research by the Department of Defense (DOD), the Office of Naval Research (ONR), and other government organizations. Two of its research centers, the Instrumentation Laboratory and the Lincoln Laboratory, accounted for over 50 percent of MIT's total budget. Moreover the two labs had almost no undergraduate function and staffed the majority of positions with scientists and engineers from outside the normal faculty.[17]

By the late 1960s Cambridge also was a central node of the global student movement. Because of MIT's relatively apolitical culture and its administrative tactics, it had maintained order through the chaotic year of 1968. On March 4, 1969, however, students and faculty shut down the Institute in a symbolic work stoppage. Over fourteen hundred students, faculty, community visitors, and reporters attended a teach-in that featured panels on sponsored research, military technology, and the Vietnam War. After the strike MIT's president Howard Johnson appointed faculty and students to a committee known as the Pounds Panel to investigate converting the Instrumentation and Lincoln Laboratories to civilian research. It concluded that weapons research made MIT "a key member of the military-industrial-university ménage à trois."[18]

MIT's shutdown was presented not simply as a question of sponsored research but as a battle over conflicting futures. At stake was the ability for scientists and engineers to speak collectively, how to redirect technical manpower from military to domestic projects, and the underlying values of the technological society. Speakers at the teach-in included Nobel Prize winning scientists, graduate students, SDS Weathermen, Congressmen, Berkeley sociologists, radical clergymen, and RAND Corporation advisers to the nation's top generals.

Though virtually every political position was represented at March 4, its impetus came from an alliance between two groups of disaffected practitioners. The first was the Union of Concerned Scientists, composed of distinguished faculty who wanted to organize the nation's scientists and engineers toward a "more responsible exploitation of scientific knowledge."[19] The UCS spoke as scientists, defending the "beneficial effects of the scientific revolution" and decrying its "misuse" as they argued for arms control hearings.[20] The second force behind the work stoppage was the student organized Science Action Coordinating Committee (SACC), founded by Joel Feigenbaum, a radicalized Cornell physics graduate student. Feigenbaum identified the "contradictions" of advanced industrial society, attached the "commingling of technology and death," and dismissed the notion that to "serve" was a higher calling than the need to "analyze and criticize."[21]

Because of MIT's status as science and technology incarnate and an extensive UCS media campaign, the March 4 protests became a national symbol of scientists in revolt. Three weeks *before* the event, the conservative columnist William F. Buckley Jr. accused the participants of "methodological anarchy" and declared their dissent to be "antidemocratic" and "seditious."[22] Commenting on the coordination of similar stoppages at thirty other universities, *Time Magazine* highlighted a sixteen-hour "Work-In" counteraction at Argonne National Laboratories.[23] March 4 was especially contentious in the technical press. The April 1969 issue of *IEEE Spectrum*, for example, reprinted the UCS' foundational manifesto as a letter to the editor. It resulted in demands that the editor resign for the "improper intrusion of politics," a member's phone call to a Congressman that he be "tried for treason," a statement from IEEE's president declaring the society would create policies for "controversial material," and a special issue giving the makers of the Antiballistic Missile program a forum to defend themselves.[24]

Engineers were not a major presence at MIT's work stoppage, accounting for only four of its twenty-six speakers. In the morality tales used to interpret March 4, however, a small group of engineers in MIT's Fluid Mechanics Laboratory played an especially prominent role. Ronald Probstein, the Lab's director chaired a panel on "Reconversion and Nonmilitary Research Opportunities," made up of engineers who implored the audience not to "sink into the system" but not to "drop out either."[25] Fluid Mechanics was touted as a potential salvation for the nation's problems by the committee that decided the fate of the special laboratories; by MIT's president; by the aerospace industry, which feared a loss of federal contracts as the nation's priorities shifted; and by the US Congress, which struggled to balance demands of the budget, the environment, industry, labor, and the antiwar movement.

Like many interdisciplinary research groups at MIT, Fluid Mechanics was an extension of a traditional department, in its case Mechanical Engineering. Also, like many laboratories at MIT, Fluid Mechanics received the majority of its funding for military applications. Until 1967 it was dedicated to research on jet engine and re-entry physics for spacecraft and ballistic missiles. Of its graduates that did not go onto similar university work, 80 percent found employment in the defense industry.[26]

Toward the end of 1966 the Lab's members resolved to convert the bulk of their research to "socially oriented" problems. At the time Fluid Mechanics consisted of three full professors, three assistant professors, a support staff, and approximately twenty graduate students. After much discussion the Lab targeted four research fields: air pollution, water pollution, biomedicine, and desalination. James A. Fay, the author of a textbook of molecular thermodynamics and a mentee of Manhattan Project physicist Hans Bethe, headed the air and water pollution group along with James Keck, an atomic weapons physicist. Ascher H. Shapiro, department head of Mechanical Engineering, went from work on jet engines to biomedicine. Probstein—a fellow of the AIAA and a PhD graduate of Princeton's Department of Aeronautical Engineering—chose desalination. None had any experience in the new areas.

Three years later, as the Lincoln and Instrumentation Laboratories came under attack, Fluid Mechanics had increased in personnel and funding. Ten projects, two-thirds of its total budget, involved converted work. By 1971 it had seventeen contracts in converted areas and an annual expenditure of $720,000. Twenty-five graduate students had been trained, completing theses on "Oil Slick Spreading," "CO Removal from Air," and "Turbulent Plume Dynamics."[27] Fay was appointed chairman of Boston's Air Pollution Control Commission, and other members took similar advisory positions. Members of Fluid Mechanics continue to work in environmental fields for the remainder of their careers.[28]

With relative ease the Laboratory translated the mathematical tools of re-entry physics and chemical kinetics into problems of pollution control and desalination. This was possible largely because of the theoretical nature of engineering science, which made conversion easier than in design and manufacturing. In the preface of his textbook *Hypersonic Flow Theory*, written before the conversion, Probstein was explicit about the theoretical nature of his work, presenting "without apology theories which are correct but which cannot be applied accurately to hypersonic flows encountered in practice."[29] The only details revealing the applications of the theory was a sentence in the introduction and a picture demonstrating flow using a silhouette of a missile.[30] The Laboratory's postconversion work similarly emphasized theory; its investigations of oil slick spreading, for example, were determined principally by simulation rather than by data collected in the field.[31]

Members of Fluid Mechanics also avoided making broad claims about technology & society. In their awakening they did not use the language of "democracy," "human values," or "the military industrial complex." They spoke instead of a personal, apolitical "reconverting our own thinking." Shapiro, for example, described his shift to biomedicine as motivated by a "latent desire from childhood to be a doctor."[32] As a group the Laboratory moreover never repudiated federal funding or military research. Probstein declared at the March 4 meetings and later in the journal *Astronautics and Aeronautics* that "it was not our intention either three years ago or now to sever relations with defense-supported research. Rather we made *an effort to redress an imbalance.*" He identified the ONR as particularly "farseeing" and the DOD as "sensitive and intelligent" in its approach.[33]

Praise of the DOD brought the Laboratory positive coverage in the technical press by offering rhetorical fodder against technology's critics.

Astronautics and Aeronautics, for example, used the Laboratory to editorialize about the aerospace industry's ability to aid in domestic problems while maintaining defense and space projects.[34] Nonetheless, Probstein criticized corporations and funding agencies that were unwilling to change, arguing that energy companies spent money advertising their commitment "to eliminate smog and to clean the oil off the birds" while refusing in practice to support such research.[35]

To scale up R&D similar to that of the Fluid Mechanics Laboratory, Probstein suggested a federal advisory committee on conversion and the environment. Government intervention was necessary because of a "monopoly of interests" detrimental to the public good, and because technical changes coupled with administrative changes would compel existing agencies that were combating urban and environmental problems with "nineteenth-century engineering" to turn to engineering scientists.[36] He was critical of the "regulatory" rather than "directly participatory" goal of the EPA and lamented that existing agencies responsible for pollution control used antiquated methods. In a congressional hearing he emphasized the primacy of aerospace and electronics industries in solving environmental problems, while lamenting that it was those projects that were being cut. Despite the naysayers, conversion would work because the aerospace firms had already shown their flexibility by *creating* the military-industrial complex. Employees who entered the workforce as computer scientists, electrical engineers, materials engineers, and the like, had developed lunar landers and supersonic transports without being formally trained to do so. He concluded that an imaginative "reconversion" of aerospace engineers would be rapid and provide immediate benefit.[37]

Back at home, members of the Fluid Mechanics Laboratory acknowledged the need for federal solutions to reduce the defense budget and tackle domestic socio-technical problems but adopted a strategy that maximized the benefits for MIT students. The Lab members' pragmatism was evident in their view of the Instrumentation and Lincoln Laboratories. Shapiro argued that the laboratories should be divested because they had only "fringe connections" with MIT's educational mission and they created negative publicity for the Institute. Shapiro, however, was against converting the laboratories to problems of urban renewal or pollution control. He argued that MIT should take a lead in a "national reordering of priorities," but that its contribution would be "superficial in comparison to [its] capabilities" if it made the special laboratories the primary site of conversion. Small, diversified labs offered more flexible opportunities. Furthermore forced conversion would be "an ineffective way of working towards reduction of the power of the military," because the Pentagon would simply redirect its funds elsewhere. "It is both futile and wrong," he claimed, "to use MIT as a club with which to beat upon the Pentagon."[38]

The Fluid Mechanics Laboratory, in short, recast its service without naming the system. Its initial conversion was done without fanfare, and members of the Lab were careful to not to fault the military or government. At March 4 and beyond, this strategy proved an effective means of generating financial and institutional support as well as extensive media recognition. But the Laboratory's local approach to national problems of conversion and out-of-control technology was not without critics. In response to Probstein's congressional testimony Murray Weidenbaum, assistant secretary of the treasury for economic policy, was skeptical of a broader aerospace conversion, citing a "chamber of horrors" of the industry's prior false promises.[39] At MIT, SACC activists rejected the notion of "working from within the system" while still receiving defense contracts. One student asked Probstein "How many Vietnamese children have died because of that 35 percent?" Another argued that the few hundred thousand dollars the Fluid Mechanics Laboratory received for pollution control research paled in comparison to the $77 million spent daily on the Vietnam War. Probstein offered his thanks for the feedback but declined to enter the fray.[40]

The Dispossessed

In the early morning of April 13, 1967, a contingent of 250 Princeton students, faculty, and locals packed into chartered buses and headed for Central Park to join 400,000 others in the Spring Mobilization to End the War in Vietnam. Prior to their departure, amid songs by the rockstar/ physics graduate student Evariste, Steve Slaby, chairman of the ad hoc student-faculty committee declared: "We hope to show that it is no longer necessary to have a sign reading 'Even Princeton' at a march, that Princeton is taking a lead and not following."[41] Even stranger than a rockstar/physicist or the claim that Princeton would take the lead in the student revolution, was that its organizer was a professor of engineering graphics.

With its reputation as a small college for the nation's elite, Princeton seems an unlikely environment for a social movement led by engineers. While it may have appeared a bastion of conservatism compared to Berkeley or Columbia, structural changes at Princeton in the wake of World War II were dramatic. The university was transformed from a school dedicated primarily to the humanities to nationally a leader in science and engineering.[42] The expansion in physics was especially notable as faculty commuted in the nation's service between suburban New Jersey, Los Alamos, and Oak Ridge. But Princeton's previously unexceptional School of Engineering also became an elite program.[43]

Engineering always had a tenuous existence at Princeton. The first courses were taught in 1875 in a university wary of professional training. Yet with the founding of a School of Engineering in 1921, accompanied by a campaign entitled "Engineering Plus," it had assimilated itself to Princeton's mission as a moral and intellectual training ground for America's leaders. Engineering remained a small but respectable presence until World War II, when its ranks swelled through the Engineering, Science, Management and War Training Program, which taught 3,619 students under graphics professor Frank Heacock. By 1945 enrollment had shrunk back to a prewar level of sixty-two students.[44]

But postwar engineering at Princeton was different in character. Government and corporate grants rolled in, doctoral study became the norm, research replaced teaching as the faculty's primary responsibility, and the university became a pioneer of engineering science.[45] At the heart of the transformation was the creation of the Department of Aeronautical Engineering. Established in 1942, and heavily supported by the Guggenheim Foundation, chairman Daniel Sayre built one of the world's premier programs by recruiting the helicopter engineer Alexander Nikolsky from Sikorsky Aircraft and Courtland Perkins from the Army Air Force. In 1950, amid rapid growth, the entire Aeronautical Engineering faculty moved off-campus to the new 825-acre Forrestal Research Center. Then, in 1964, Aeronautical Engineering absorbed the faltering Department of Mechanical Engineering to form the Department of Aerospace and Mechanical Sciences.[46]

Not all engineering faculty at Princeton, however, benefited from the boom in federal funding and the rise of engineering science. Those in disciplines not oriented toward defense research found themselves on the margins. Civil engineers in particular scrambled to make alliances. They contended that their main partnership was with the social sciences, which they argued made them the mediators "most useful to society." Even as they spoke of assimilation, however, Princeton's civil engineers admitted that they worked on the "fringes with the pure sciences."[47]

While civil engineering merely was at risk, the Department of Graphics and Engineering Drawing was deemed obsolete. But graphics, which traced its roots to the origin of engineering at Princeton, would not go quietly. In 1953 the department hired associate professor Steve Slaby to help resurrect its waning program. Born in Detroit in 1922, Slaby had an atypical education for a Princeton professor. For one thing, he did not have a PhD; he received his undergraduate training in mechanical engineering at the Lawrence Institute of Technology, a small college adjacent to Ford's Highland Park factory, and his master's in economics from Wayne State University. During World War II he was an aviation cadet. He then taught engineering graphics at Sampson College, a short-lived New York college designed to teach veterans on the GI Bill. Slaby moreover was an avowed socialist with experience in labor relations gained on a Fulbright fellowship in Norway.[48]

The seeds of Princeton's antiwar movement began in a 1956 dispute over the engineering curriculum. Vital to the School's move toward engineering science was the revision of its common freshman year.[49] As it stood, graphics was the only required part of every freshman's coursework, with a semester each of engineering drawing and geometry. Because the Graphics Department did not grant degrees and lacked upper level courses, its survival hinged upon its service role. Its first setback came when electrical engineering faculty suggested that drawing and geometry be compressed into a single course. Slaby argued that graphics was foundational to all engineering fields because it enabled the engineer to "bridge the gap between the symbolic or mathematical with the physical or real."[50] His critics, however, characterized "drafting" as a task for the technician, rather than the scientific engineer. Slaby raged that the "language" of graphics was a source of visual communication and creativity and that the purpose of engineering was to merge the human with the technical.[51] Despite his objections, Slaby was forced to combine the courses. Graphics suffered its final blow when the School made his course a pure elective. Enrollment plummeted. In the course's last offering, Slaby had only 2 of the School's 171 students.[52]

This struggle over pedagogy and engineering philosophy might be interpreted as typical of postwar tensions within the academy, as the rise of new methodologies opened new fields at the expense of those deemed outdated. But Slaby's framing of his technical marginalization became a catalyst for political activism. He contended that graphics' reduction would have a detrimental impact on student formation, distributing statements from engineering dropouts to prove there was an "engineering school problem" caused by the excessive focus on engineering science, which at root, was driven by military applications.[53] He highlighted on one of the proposed replacements for graphics, a course that would instill "the value of basic mathematics and science to the engineer" by exploring nuclear energy, detection and destruction of ICBMs, and guidance systems for missiles and submarines. He implied that these "negative" applications were part of a systematic expansion of government and military funding embodied in the engineering-science philosophy.[54]

Throughout the 1960s Slaby developed a vision of engineering service that rejected the ethos of technology as inherently "for the progressive well-being of mankind." He argued that students needed to be trained to recognize the politics of engineering knowledge, and as early as 1961 he and the civil engineer Sumner B. Irish unsuccessfully proposed a required freshman technology & society course. They stressed the moral and political potential of "service to mankind" through civil engineering, while at the same time asserting that "the greatest threats" came from "physicists, electrical, mechanical, aeronautical engineers who today derive a tremendous amount of their support from military connected projects."[55]

As his graphics obligation deteriorated, Slaby retooled as a politics instructor and dedicated himself to campus organizing. He became one of the leading advisors to the student movement as well as an avid interpreter of theorists of technological politics. Slaby's own writing from the mid-1960s into the 1970s highlighted what he viewed as the cage-like environment of modern engineering. He worried that resistance was useless and confessed to the "feeling of being trapped." Echoing Marcuse, he described the "unfreedom" of "mind and soul" as "most deadly in modern so-called capitalist affluent and materialistic society."[56]

In the fall of 1967 the role of military-sponsored research emerged as the most explosive issue on campus when an article in the student newspaper drew attention to a branch of the Institute for Defense Analyses (IDA) at Princeton. The IDA was a private think tank for the DOD nominally independent of Princeton but located in a building adjacent to the Engineering Quad.[57] SDS staged a sit-in and a hundred faculty members petitioned for severing ties with IDA.

In the aftermath of the IDA affair, Slaby mounted a relentless campaign against defense research at Princeton. The IDA was an obvious target for students, but sponsored faculty research was a less tangible reality.[58] It was all too familiar for Slaby; "after all," he wrote, "how can one who is receiving major support from the Air Force talk against bombing in Viet Nam."[59] To bring the extent of the problem to light, he called for a public bibliography of all funding on campus.[60]

Events at Princeton reached a climax in the spring of 1970, when President Nixon announced the expansion of the war into Cambodia. In the university chapel, students demanded a shutdown. Slaby rallied in support: "if the situation doesn't change," he said, "we may have to close down the university for a few days, for a few months, until it does."[61] Early the next morning a small group of students gathered at the Mather Sundial. Slaby and religion professor Malcolm Diamond accompanied their classes to the Dial, bringing the crowd to a hundred. By day's end, there were a thousand. After what one student described as "an odd mixture of intensity and—well, two weeks at summer camp," the student movement wound to its conclusion. The term ended. The students went home.[62]

Princeton's strike prompted an institutional reevaluation of sponsored research. Historian of science Thomas Kuhn chaired a committee that examined classified research at thirty-four universities. It painted Princeton in a relatively positive light compared to peer institutions, but Engineering School faculty came out of the committee chastened. Hardest hit were the aerospace engineers, who made the migration back to the main campus from their special laboratory in a newly renamed Department of Mechanical and Aerospace Engineering.

The Kuhn Committee gave Slaby a measure of vindication. In the atmosphere of accommodation, the technology & society course he had lobbied for nine years earlier was approved and became a staple undergraduate offering. Throughout the 1970s students read technology's critics and worked in teams on water pollution studies, waste management, public housing, and decentralized communications systems.[63]

Long after the events of the 1960s Slaby remained committed to institutional social justice. Unlike the world of private industry, where to be an agitator often meant the end of one's career, tenure provided Slaby the opportunity to reinvent himself as a socio-technologist in a space where he had to be tolerated. "Only if academicians and scholars can disengage themselves from [their] servant status," he would argue throughout his career, "is there a chance that they can acquire that independence which is crucial to true free inquiry, critical analysis, and wisdom." Once engineers' intellectual freedom was assured, they could be "transformed from servants and bystanders to active participants in the making of history."[64]

Responsibility beyond Professionalism?

The Committee for Social Responsibility in Engineering had the tools to reveal the false promises of cold war service, if competing strategies for remaking their profession. To understand how engineers cast overt politics as a national project that could cut across academe, industry, government, and the member societies requires returning to the streets of Manhattan were this book began. There, in March 1971, the CSRE announced itself to the world at the annual convention of the IEEE. Dissident engineers recited a litany of failures. The prosperity of the technological society had been built on "doing unnecessary work, on socially non-productive projects." At the same time the IEEE had failed to "serve the needs of its membership" and instead produced "meaningless gestures" that were an "affront to its members."[65] Faced with unemployment, misused talents, and weapons technology that "spell[ed] ultimate disaster for mankind," the CSRE set as its purpose to "challenge the present orientation of engineering and to explore ways in which engineering skills can be used to solve the obvious and growing ills of our society."[66]

The CSRE was born out of friendships among antiwar radicals, veterans of the scientists' movement, and senior researchers at Bell Laboratories, Brookhaven National Labs (BNL), Brooklyn Polytechnic Institute, Columbia University, IBM, New York University, and RCA. Its "sparkplug" was Ted Werntz, a pro–civil rights, antiwar, anticorporate nuclear engineer at BNL.[67] Brad Lyttle, the director of the War Tax Resistance, provided insight as a full-time political organizer and an engineering dropout.[68] Stephen Unger, a young electrical engineering professor at Columbia, also played a formative role. Unger's research focused on designing electronic networks for community interaction. His co-principal investigator, the sociologist Amitai Etzioni, described their goal as the design of electronic networks to re-create participatory democracy akin to the ancient Greek *polis* and the Israeli *Kibbutzim*.[69]

Though the CSRE chose to work outside of the IEEE, the professional member societies gave the organization its raison d'être. CSRE members had been rebuffed in their efforts to change the IEEE from within, which they took to task for "evading" politics. The CSRE called the member societies' position of "neutrality" a "façade" for control by top management, whom it tied to misdirected technological imperatives and documented how most IEEE fellows (the organization's highest distinction) had worked on defense research.[70] Still most CSRE members were attempting to democratize the IEEE with the goal of making engineers more effective professionals. Unger and others recognized the engineer as an "employee-professional" and called for an industry tenure system to create an atmosphere conducive to the "responsible professional."[71] The CSRE first petitioned the Executive Council to establish a "Professional Group on Social Implications of Technology," and later to define the IEEE's primary mission as the promotion of engineers' "economic well-being."[72]

Anyone who picked up the CSRE's journal *Spark*, however, could see that it was a window to alternative visions of engineering that went beyond professionalism as an organizing ideal. Its cover greeted the reader with a clenched fist, from which a lightning bolt shot out toward a transistor to form a peace sign. Inside, they learned about "doing it at the workplace" and the "electronic control of deviant, dissident, and delinquent Americans."[73]

Figure 5.1
Spark
Source: *Spark* 1, no. 1 (spring 1971): cover. Courtesy of Stephen H. Unger.

Instead of creating a public interest organization designed to preserve an existing identity, CSRE members experimented with a spectrum of possibilities for engineering service. *Spark* introduced readers to virtually every alternative science and technology organization of the day. Its first issue consisted of articles, book reviews, and ephemera produced by thirty-nine other groups. By functioning as a clearinghouse of information, the CSRE sought to decrease "isolationism" fostered by the military-industrial complex and convince engineers that dissent had national momentum. A range of organizations advertised in *Spark* with the goal of converting military service to public service. These included national groups like the Scientists' Institute for Public Information and the Federation of American Scientists to small companies such as Pacifica Engineering, a New York firm with the mantra "Creative Designs for the Common Good."[74]

In their pitches to *Spark* readers, alternative organizations emphasized the process of critical awakening. The California-based Employment Clearing House, a nonprofit agency for relocating defense industry employees into civilian jobs, for instance, was formed out of the Technology and Society Committee, a Palo Alto salon concerned with the "antipolitics" of science and technology, that itself had emerged from lunchroom discussions of Marcuse and Ivan Illich at the Stanford Linear Accelerator.[75] The CSRE stressed a similar strategy, recommending that concerned engineers start lunch discussion groups at work.

Contributors analyzed the false consciousness of professionalism in corporate employment, explaining how the system created images of mobility that acclimated individuals to its goals:

We worked, studied and got ahead. After graduating from a demanding technical program, we were off to a good job at high pay and, who knows, maybe a shot at middle management. Before us flowered the prospect of a professional career, status, upward mobility and exciting technical work pushing forward the "state of the art," not to mention an occupational deferment. This was the optimistic hope.[76]

Instead of a confirmation of that hope engineers struggled to maintain employment under terms dictated by project-based defense contracts. Only when science and technology were put in the "direct control of the people" would their alienation cease to exist. Lee Horowitz of the Newark College of Engineering likewise highlighted the structural inequality of the technological society:

I applaud the fighting spirit shown by those who hope to make the IEEE into an instrument in the service of the working engineers, but I gravely doubt whether such an attempt is either possible or, in the long run, desirable. . . .

Furthermore, we can hardly claim that a new exclusionary organization for the betterment of those who happen to be engineers is any greater boon to mankind than a racist construction union or an elitist medical guild. The problem is not to improve our relative position within an anachronistic and barbaric system, but to abolish the system.[77]

For most dissidents, social theoretical texts were useful for naming the existing system but did not constitute an encompassing ideology. A handful of organizations, however, hoped to make an ideology of technological politics an engineering philosophy. The Berkeley-based Aquarius Project, for instance, sought to recruit *Spark's* readers to pursue "Revolutionary Engineering" for the design of "Counter-Technology." The organization believed that the movement needed people with technical backgrounds as much as engineers needed the movement. Engineers would gain social consciousness while the counterculture would gain a deeper understanding about technology. In the service of intentional communities built from the detritus of "waste-capital," technology could serve the people as technical and nontechnical participants repurposed it in tandem.[78]

In contrast to moral individualist, new Left, and countercultural strands of engineering thought, unemployed engineers turned to *Spark* seeking a catalyst for organized labor. Unions recruited new members by blending technical conversion with traditional labor politics. The Council of Engineers and Scientists Organizations (CESO), an association of 23 engineering unions representing 125,000 employees of the largest defense contractors, argued for the creation of a NASA-style federal agency for pollution, mass transportation, and crime. The CESO, however, also lobbied Congress to take an isolationist position on immigration, protesting policies designed to recruit foreign engineers.[79]

Testimonials from engineers were critical for giving a human face to protest and alternative service. Letters from corporate dropouts reinforced the argument that technical work could be based on moral conviction. The converted recounted how they had tried to forge careers without working on weapons design, but found that companies applied their civilian research to military applications. They often emphasized the importance of technology & society literature in their transformation.

Others related the struggle back to the CSRE. One reader recounted his pride in provoking uproar among "conservative" and "fascist" engineers by placing copies of *Spark* in his company lounge.[80] Dispatches outlined the promise and peril of radicalism. Honeywell engineer Denis Brasket described the economic risk involved: ostracizism by co-workers, the inability to perform "non-military" work in corporate engineering, the recognition that one could "leave only once" and that once you did, you would lose the power and privilege afforded by being on the inside. At the same time he conveyed the rush of freedom that came from righteous action.[81]

By fusing moral responsibility with economic survival, Melman's analysis of the military-industrial complex provided the CSRE with the closest thing to a unifying theory. As its target moved beyond the IEEE, the CSRE sought to raise consciousness about engineers' complicity in the military-industrial complex. Members used their "inside" position to promote a slide show on the National Action/Research on the Military-Industrial Complex that explained the technical details of the automated air war in Vietnam. It lent out the film and projection equipment to interested parties and offered screenings at technical conferences and corporate headquarters.[82] The CSRE's technical expertise also generated frequent calls for assistance from activist groups such as the Honeywell Project and the GE Project that targeted corporations for their defense contracts.

The CSRE successfully created a space for engaging radical politics and maintaining an identity as engineers, but the experience was short-lived, with the last issue of *Spark* published in 1975. The CSRE's mercurial existence was in part due to personal conflicts but it was also the result of its open-tent philosophy and the friction between intellectual and economic motivations. Werntz and fellow CSRE member Paul Stoller, for example, accused unions of being co-opted by employers, discriminatory, and supportive of the military-industrial complex.[83] Additionally the United States' withdrawal from Vietnam and the bottoming out of the recession diminished the sense of revolution.

Above all other factors in the CSRE's demise, however, was its success in forcing changes in the IEEE, which split those who saw professionalism as the best means to public service from those who believed only opposition could bring change. In 1972 the IEEE rejected the CSRE's petition for a technology & society group with published transactions and local chapters but approved an ad hoc Committee on Social Implications of Technology (CSIT). The Committee, which distributed its first newsletter in December 1972, resembled Paschkis' group in the ASME. Most of the CSIT's organizers were CSRE participants who shifted their energy to the new IEEE organization. The CSIT's constituency was more interested in ethical codes, standards, and using technological systems to solve domestic social problems than they were in attacking the "military-industrial complex." While not beholden to a single doctrine, the CSIT worked to address technology's social implications by exploring how corporate employees could be ethical professionals.

Where *Spark* opened its pages to the radical left, the CSIT's newsletter hewed toward the center. When participants brought social theory into the discussion, their sources were more likely to be Daniel Bell and Alvin Toffler than Illich and Marcuse. The CSIT surveyed developments in engineering education and reprinted speeches by center-left elites such as MIT president Jerome B. Wiesner, who argued that it had been no one's responsibility to control the pace and scale of technological change, but now technology's unintended effects demanded regulatory action.[84]

Nonetheless, the CSRE's activist origins infused the CSIT with a sense of service as loyal opposition. For example, the group organized an IEEE-sanctioned panel on "The Engineer and Military Technology" at the 1973 Convention that brought together a deputy undersecretary of the Air Force with a pacifist physician once jailed for refusing to advise the Special Forces and a member of the IEEE's Board of Directors who argued that engineers were responsible for serving society regardless of whether its goals were curing cancer or designing weapons. The CSIT also argued for the abolishment of events requiring security clearances. Instead of pillorying the IEEE in a "disruption exercise," however, the CSIT raised the issue civilly in formal channels.[85]

Unger was the CSIT's motive force. A friend and colleague of Paschkis's at Columbia, Unger took up professional ethics as a means of strengthening engineers' right to moral judgment in corporate work. Traces of the CSRE's activism can be seen in his effort to revise the ECPD's canons of ethics. Unger's rewriting reflected the CSRE's antiwar position—declaring that an engineers' primary responsibility was "Using his knowledge and skill for the advancement, *never the detriment*, of human welfare." Subsequent canons attempted to assure that an engineer make "a reasonable effort to inform himself as to the possible consequences, direct and indirect, immediate and remote, of projects he is working on" and educate the public about the multiple "alternatives offered by modern technology."[86]

Unger and the CSIT found their cause célèbre in the case of San Francisco's Bay Area Rapid Transit System (BART). This was a controversy that brought to bear a classic conflict between employees and employers that also spotlighted the aerospace industry's foray into urban development. Three engineers had been fired in 1972 for circumventing their management when they encountered what they believed to be inadequate safety systems for the billion-dollar automated transit project. They appeared to be vindicated when a test train over-ran a station. Unger used the CSIT as an interpreter and amplifier of the BART controversy, emphasizing the challenges faced by the California Society of Professional Engineers in their defense of the fired engineers. The CSIT pushed the IEEE to action, and in a landmark decision for a member society, the IEEE filed an amicus curiae brief on behalf of the terminated engineers that characterized BART's actions as a breach of contract.[87]

An emphasis on alternatives and loyal opposition helped the CSIT thrive in the long-term compared with efforts in other member societies or with the provocative language and tactics of the CSRE. Despite championing its openness to all IEEE members, the CSIT remained controversial because it challenged the notion that "existing technology affects social systems and, conversely, societies effect their own technological development."[88] In 1981, however, by a narrow margin it achieved the society-level status it had petitioned for nearly a decade earlier, giving the new IEEE Society on Social Implications of Technology the ability to print official transactions in the quarterly *IEEE Technology and Society Magazine*. A recent issue includes a feature article on the potential implications of nanotechnology in Iran, the absence of privacy in a digital society, and an analysis of engineering ethics.[89]

Conclusion

In the early 1970s it was hard to find anyone arguing against a partial conversion of America's technical workforce from defense projects to domestic ends. Aerospace industry managers met in national conventions with the US Department of Housing and Urban Development. The NSF initiated the Research Applied to National Needs (RANN) program. Simon Ramo—who diversified TRW's clients to include the City of Los Angeles—championed the "social-industrial complex" in the *Wall Street Journal*.[90] Corporate and government diversification was reinforced by a narrative claiming that retraining the nation's scientists and engineers would alleviate unemployment, reinvigorate a service mentality among professionals, and restore the public's faith in technology.

The ease with which the nation's largest defense contractors embraced conversion narratives helps explain why dissidents saw naming the system as vital. "The military-industrial complex" conjured almost universally negative connotations that could be answered only by claims that it did not exist or that its leading figures were best situated to transform it.[91] Those who turned to the typewriter, the printing press, and the megaphone emphasized distinctions between "real contributions" and a "thinly disguised attempt to channel funds into the aerospace industries." In a flier distributed at the Urban Technology Conference—an event co-sponsored by the AIAA and the National League of Cities—the CSRE urged support for "a farsighted definition of goals which include considerations of environmental balances and material resources," but railed against "expecting quick cheap technical fixes to deep rooted social problems; arrogant top-down decision making without community participation; . . . [and] the use of urban studies contracts to tide over weapons companies between military contracts."[92]

For dissenting engineers, appropriating the symbols and tactics of movement politics was a persuasive strategy that drew media attention and conveyed moral conviction.[93] The CSRE counterconference resulted in generally favorable press due to the surprise of engineering activism. But this was sustainable only insofar as engineers stayed within accepted bounds of civility and professional norms. In 1971 the CSRE adopted a teach-in format and quietly protested keynote events with placards, but maintained deference to the speakers. This prompted the journal *Electronics* to echo the sentiment of an IEEE convention organizer that the protesters were "good-quality dissidents."[94] When the CSRE turned to more disruptive tactics at the 1973 convention—teaming up with SftP and other protest groups to stage a "Computer Theater" outside the New

York Coliseum—they were pelted with paint bombs and eggs from inside the convention.[95]

Defenders of the aerospace establishment also saw the appeal of a radical affect. In 1970 University of Michigan aerospace student David Fradin and his professor Wilbur Nelson created the student group Fly America's Supersonic Transport (FASST) to combat technology's critics.[96] The novelty of a pro-technology student group propelled Fradin to congressional testimony on the fate of the billion-dollar project. Rather than disbanding when the supersonic transport was terminated, FASST rebranded as the Federation of Americans Supporting Science and Technology. It described its mission as the creation of a harmonious "spaceship earth." Emulating the underground press, *FASST News* printed testimonials from engineers and politicians that explained the enthusiasm of aerospace's pioneering days in the 1950s, cast "gutless critics" as the source of America's malaise, and described a path to the restoration of technology's glory.[97] Unlike the left-leaning groups it emulated, FASST's national advisory board included the defense intellectuals Paul McCracken and General Bernard Adolph Schriever; its legislative advisors were Barry Goldwater Jr., Jack Kemp, and other conservative politicians; and it received support from nearly fifty corporate backers. FASST initiated chapters at universities and high schools, compiled a narrated slide show titled "Technology: Friend or Foe," and distributed smiley face buttons whose teeth spelled "Think Space Shuttle."

Whether championed from the left or the right, widespread conversion of military research and development to "socially responsible" applications, though a subject of multimillion-dollar experimentation and much publicity, did not amount to a drastic rearrangement of national priorities. Engineers did not withdraw their services from defense industries in significant numbers. They certainly did not tear the "racist, imperialist system to shreds."[98]

While public relations campaigns spawned in response to activists contributed to the marginalization of overt politics in engineering, struggles to scale up alternative visions of technology were also a consequence of the challenges of self-definition. An ideology of technological politics compelled a withdrawal from the professional channels of science and engineering. As engineers' testimonials attested, this was a transformative but often binary decision that involved shedding friendships and work

"THINK SPACE SHUTTLE " BUTTONS

FASST now has a limited supply of these Space
Shuttle buttons for distribution to members.
They're free, so get yours while they last!

Figure 5.2
FASST buttons
Source: "Think Space Shuttle," *FASST News* (July 1973). Courtesy of David Fradin.

relations. Abandoning deeply held vocational ideals was a choice most were unwilling to make. In her study of out-of-work technologists on Route 128, sociologist Paula Goldman Leventman was surprised by their support for antiwar candidate George McGovern, combined with a general silence about the structural causes of their unemployment. "Seventies professionals did not want to change the political system," she concluded, "Unemployment sometimes stimulated insight but the very process of politicization threatened the generalized status orientation of their total life situation."[99]

The gap between the accepted states of activist and engineer—and the trade-offs of either state—is best seen by comparing the lives of two engineers who passed through the same field on divergent trajectories. Antiwar politics had a profound effect on Unger, one of the CSRE's founding members. In 1972 he seized the opportunity afforded by the CSIT to champion his vision inside the IEEE. He later published the textbook *Controlling Technology: Ethics and the Responsible Engineer,* which offered engineers a third way between doomsayers of unstoppable momentum and supporters of inevitable adjustment. In Unger's view, engineers were not technology's masters (nor even its decision makers), but because they proffered expert advice and turned social vision into reality, they had an outsized societal impact. Closest to the machine, they were the first to recognize when things went wrong. Most important, engineers' choices resulted in dramatically different futures. Given these conditions, the way to an equitable society was to enhance professional awareness of how technical choices were made, combined with provisions for sanctioned dissent.[100]

The disillusion of cold war service also changed Chris Murray's life. A "nice engineering-type girl" at GE, Murray wrote to her colleagues, her bosses, and *Spark's* readers that defense research had driven her to "an alternate mode of living" as an organizer for the Syracuse Peace Council. At first she had protested the Vietnam War by putting up posters on her office walls, wearing an armband to work, and attending demonstrations. Such actions "marked a difference" between her and colleagues, but also generated discomfort among activists. A choice had to be made. Though she cherished her working relationships and acknowledged there was "no simple solution" to the problems of technical conversion, for her, control could only be found "working on the outside of these monster corporations."[101] Her manager tried to convince her to stay, admitting that he himself desired to work on medical applications, but that he was obligated to pursue defense work over his personal qualms.

Engineers who viewed the "military-industrial complex" through the lens of technological politics repudiated service to the corporation, to the state, and to professional norms. In its call for a rejection of the existing system, radicalization embodied an element of release: "I realize that very few people are in the personal situation that I am," wrote Murray, "one that allows the freedom of making a full commitment to any cause."[102] But it was one thing to identify professionalism's false promises and quite another to suggest that confrontational politics could replace the judgment of experts and the security of established employment structures. Technological politics, when raised to the status of ideology, characterized the cage as containing vastly more than it excluded and made the incompatibility of being radical and being engineers self-fulfilling.

6

Three Bridges to Creative Renewal

If we cannot be inattentive to change of objectives in our society and cannot accept the silent calls of return and revolt, what is left?
—Donald A. Schön, *Technology and Change: The New Heraclitus*[1]

Long before engineers became embroiled in debates about out-of-control technology, the novelist Kurt Vonnegut Jr. predicted a key feature of their response. In his 1952 masterpiece *Player Piano*, the former GE technical writer described an over-rationalized world partitioned into communities of engineers, machines, and everyone else. A technological revolution based on automation and meritocratic sorting had fulfilled the material needs of society but had banished the value of work with "clever . . . hands," generating a pervading sense of uselessness among the mass of humanity left behind. Society's new elites moreover felt "annoyed, bored, or queasy" with the system they had built. Dedicated to anonymous corporate service, engineers had loveless marriages, uncertain job security, and lacked authentic experience. Under these strains Paul Proteus, the thirty-five year-old manager of the Ilium Works—the highest paid man in the region—began to crack. Along with Edward Finnerty, a brilliant nonconformist engineer and virtuoso pianist, Proteus crossed the river that separated engineers from commoners to drink in a dingy Victorian saloon. There the two "malcontents" learned about a powder keg of resentment that portended societal revolt. Seeking a movement founded on community and the "uplift of creativity," Proteus issued a desperate plea to "meet in the middle of the bridge."[2]

In the 1960s a number of heirs to Proteus and Finnerty rallied their peers to similar reconciliatory acts. Alternatives to the technological society, they contended, would not be found in ethical codes or confrontational politics but rather through a rekindling of the adventure that drew them to engineering in the first place. In projects ranging from one-to-one aid in the developing world to partnerships with avant-garde artists and humanist intellectuals, engineers sought to make the design process itself a tool of creative renewal by emphasizing collaboration with users. Reformers with wide-ranging commitments expressed a common desire for personal and collective identifications apart from the organization man achieved through new work experiences. Most of all they posited faith in technology's "human" capacity, which served as the principal link to their nonengineering collaborators. Vested interests, social conformity, and the limits of a scientist's mentality, they argued, were to blame for the engineering profession's inability to achieve a middle path in which technology aided community growth and self-actualization.

At the end of the decade the rhetoric of humane technology reached a fever pitch in the profession. In May 1969, for example, Vice Admiral Hyman G. Rickover, designer of the nuclear submarine, lectured on the need for "humanistic technology" that would result from a "humanistic profession."[3] That same month William Leavitt, a senior editor of *Air Force/Space Digest*, described the aerospace industry's path "Toward a Humane Technology" through investment in a "decentralized, community-oriented technological attack on social problems."[4] This holistic ideal found clearest expression in Samuel Florman's 1976 *Existential Pleasures of Engineering*, published after decades of his straddling the two cultures divide. "Let us create a *new guild of craftsmen*," Florman quoted Bauhaus founder Walter Gropius, "without the class distinctions which raise an arrogant barrier between craftsman and artist. Together let us conceive and create the new building of the future."[5]

This chapter investigates engineers' ambitions to remake society and the character of their profession under the banner of humane technology. It focuses on network-building reformers who deliberately crossed boundaries between expert and user, research laboratory and artist studio, development agency and corporate boardroom. I explore three orientations of self, society, and technology as they emerged in engineers' bridge-building networks. The first investigates engineers' contributions to the appropriate technology (AT) movement in the organization Volunteers for International Technical Assistance (VITA). Considered by proponents and critics alike to be the material instantiation of an ideology of technological politics, AT has been praised as an alternative to authoritarian technology and dismissed as a movement of ideologues.[6] But, from its very origin, AT also had strong links to corporate and academic engineers who championed creativity, prudence, and service as professional ideals. The second turns to Experiments in Art and Technology (E.A.T.), a freewheeling network that sought to resolve the dislocations of technological change and achieve new professional selves through partnerships with professional artists. The third explores the Innovation Group, a collaboration of high-technology managers, journalists, and social theorists whose publications and executive salons fostered a new professional archetype, the change manager. Part technical expert, part humanitarian, and part hardheaded competitor, the change manager would be an adaptive agent in a world beyond his control.

Each network originated prior to widespread fears of out-of-control technology but was transformed by that crisis, and owed its rapid growth to it. Each peaked in the early 1970s with varying degrees of achievement and recrimination. In comparison to member society reformers and oppositional movements, each was largely successful in re-infusing technology with progressive meaning. In their vivid images of a technologically mediated revolution, however, each drew upon and contributed to the notion that *engineers* were most in danger of being left behind.

To Engineer Is Human

The notion of humane technology is a well-tread problem in the history of the 1960s. In 1959 C. P. Snow's *Two Cultures,* and the symposia it spawned, reinvigorated a perennial argument about the merits of a literary versus a scientific worldview.[7] In the mid-1960s critics of the technological society shifted the conversation from "human values" to the prospects of self-determination in an artificial world. Technological systems, they claimed, had become too large and too centralized. Access was limited to experts in rigid, impersonal institutions working toward destructive ends. But technology also could be remade to enhance personal liberty and community growth.[8]

Humane technology found its greatest advocates in a series of countercultural projects that eschewed confrontational politics for what the iconoclastic designer R. Buckminster Fuller called "the design revolution."[9] Inspired by Fuller, the media theorist Marshall McLuhan, and other optimistic futurists, participants in these projects saw technology as a tool for social harmony. E. F. Schumacher, the economist turned AT pioneer, championed "technology with a human face" in contradistinction to "super-technology" that "deprived man of the kind of work that he enjoys most, creative, useful work with hands and brains."[10] Artist collectives like USCO built communal environments with lasers and electronic music. Stewart Brand's *Whole Earth Catalog* sold technologically mediated liberation to all comers.

In championing these experiments, recent scholarship has discarded stereotypes of anti-modern hippies in favor of a new consensus view of techno-utopian "legitimacy exchanges." At elite institutions in proximity to bohemian enclaves—Stanford Research Institute, MIT, Bell Laboratories, Xerox PARC—communalists, artists, and musicians traded ideas, access to technology, and credibility with research scientists. In his study of Brand and the *Whole Earth* network, Fred Turner shows that in the "forgotten openness" of the military-industrial complex, the creative society was seen as an inevitable outcome of the digital revolution. When communal movements collapsed in the 1970s, these connections contributed to associations of computers as "personal" and the Internet as a haven for "virtual communities."[11]

Revisiting humane technology from the perspective of the engineering profession reveals a more contested picture than either neo-Luddism or the triumph of hippie entrepreneurs. To be sure, thousands of technical workers were drawn to countercultural lifestyles in search of meaningful work.[12] Like their communal counterparts, engineering reformers cast humane technology as a middle way that transcended partisan politics. But, rather than a flight from politics, ideology and identity were entangled in their conceptions of what it meant to be human. Iconoclasts and dropouts self-consciously defined technological selves outside the framework of standardized education and hierarchical labor. At the same time engineers were drawn to such projects because they evoked myths of craftsmen and tinkerers, individual makers of things. Finally, many argued that they already belonged to a humane calling. "For too long," the editor of *Chemical Engineering Education* wrote, "we have let the humanists suggest that they are the salvation of mankind and that the 'technologists' are the destroyers, the polluters, and the dehumanizing materialists."[13] The pursuit of practices steeped in "human values," "humane technology," "humanistic technology," "engineering as human," or "humane engineering" shifted beneath engineers' feet. Identifying what constituted humanness was at times a problem of aesthetics, technical design, public relations, philosophy, psychology, identity politics, or management. What it meant for engineering or technology to be human moreover drew upon varying imagined histories and presented no common path forward.

Anxieties and aspirations about humane technology in the engineering profession coalesced around two interconnected virtues—*creativity* and *collaboration*—that resonated with longstanding arguments about what counted as good engineering. Both were *process* values. Advocates spoke less about specific artifacts than a "search," a "path of discovery," or a "sociological revolution." But creativity and collaboration also were moral statements about the self in which engineers turned to an ancient past of Vitruvius and a speculative, cybernetic future to rethink professional ideals. They stressed design and innovation in contrast to engineering science as a means of instilling human values into technology, and new organizational patterns as a means of adjusting to technological change.

For engineers—as it was for artists, scientists, and intellectuals—creativity was a romantic expression of human autonomy offered in opposition to criticisms that technology was self-directed and that its practitioners were soulless.[14] The late 1950s saw the beginning of a veritable creativity boom in engineering. Against the backdrop of critiques of mass society, middle management guides to *Professional Creativity* and *Creativeness for Engineers* proliferated. J. H. McPherson, manager of Dow Chemical's Psychology Department, for example, preached that: "a creative engineer is an inventor, a species of artist. He's no organization man."[15] Mechanical engineer Harold R. Buhl of Iowa State likewise argued in *Creative Engineering Design* that the engineer "must possess constructive discontent regarding his environment, non-conformity regarding possible solutions, and an inquisitive 'why' attitude."[16]

Collaboration was a means of overcoming politics by breaking down established social hierarchies. The enthusiasm for collaborative work had multifaceted origins from a fixation on "interdisciplinarity" as the key to victory in World War II and cold war democracy in its wake to the network flows of postwar systems engineering, to basic humanist ideals of the fellowship of man.[17] Collaboration involved both giving and getting. For engineers, working with nontechnical experts embodied the notion that creative solutions to their profession's problems could not be achieved alone and that engineers could respond creatively in realms where they had not previously applied their expertise. In short, collaboration certainly involved calculated exchanges of legitimacy, but the very process of interaction often was interpreted as its own reward.

Despite reconciliatory, universalist inflections, however, notions of creativity, collaboration, and humanness were weapons of cultural politics. Engineers and their collaborators who took up the cause of humane technology encountered questions of agency and identity at the center of technology & society debates. What mix of "high" and "low" technology was appropriate? What arrangements of power and authority in the collaborative design process would achieve best results? Should the process itself be valued over intentions or outcomes? Who was being made "human," engineers, their collaborators, or the users of the resultant technology? How much could socio-technical change ever be controlled? For theorists like Mumford, Ellul, and Illich, overcoming the determinisms of the technological society required a redistribution of power that entailed a radical rethinking of expertise. Theorists of technological change argued in response that this critique was little more than a repackaged classicism and that technology was society's most important creator of *new* values.[18] In other words, the problem of humane technology was not overcoming technology's determinisms but rather designing democratic solutions to live in harmony with them.

Humanitarian Engineering

According to General Electric's public relations office, Dan Johnson had "a flair for making things."[19] In a 1968 recruitment campaign targeted at student journals from *Spartan Engineer* to *Yale Scientific*, GE posed the bespectacled electrical engineer working intently under what appeared

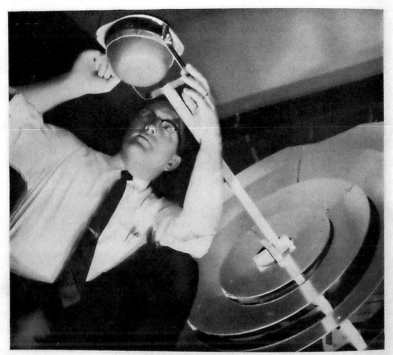

Dan Johnson has a flair
for making things.

Just ask a certain family in Marrakeck, Morocco.

A solar cooker he helped develop is now making life a little easier for them—in an area where electricity is practically unheard of.

The project was part of Dan's work with VITA (Volunteers for International Technical Assistance) which he helped found.

Dan's ideas have not always been so practical. Like the candlepowered boat he built at age 10.

But when Dan graduated as an electrical engineer from Cornell in 1955, it wasn't the future of candle-powered boats that brought him to General Electric. It was the variety of opportunity. He saw opportunities in more than 130 "small businesses" that make up General Electric. Together they make more than 200,000 different products.

At GE, Dan is working on the design for a remote control system for gas turbine powerplants. Some day it may enable his Moroccan friends to scrap their solar cooker.

Like Dan Johnson, you'll find opportunities at General Electric in R&D, design, production and technical marketing that match your qualifications and interests. Talk to our man when he visits your campus. Or write for career information to: General Electric Company, Room 801Z, 570 Lexington Avenue, New York, N. Y. 10022   (69-23)

GENERAL ELECTRIC

AN EQUAL OPPORTUNITY EMPLOYER (M/F)

Figure 6.1
"Dan Johnson has a flair for making things"
Source: General Electric Company, "Dan Johnson Has a Flair for Making Things," *Spartan Engineer* 21 no. 4 (May 1968): 26. Courtesy of GE.

to be a prototype satellite dish. But readers learned that Johnson in fact was adjusting the five-dollar solar cooker he helped design as a founding member of Volunteers for International Technical Assistance, an organization created to put technology to beneficial use in the developing world not through nation-building hydroelectric dams but rather in personal collaborations.

At the time VITA counted over 5,000 volunteers from 55 countries working on 11,000 aid requests. It had distributed 20,000 copies of its *Village Technology Handbook*, a "how-to" collection of skills ranging from stove design to hydraulic flow.[20] With funding from the US Office of Economic Opportunity (OEO), VITA soon ventured from Africa and Latin America to inner city Boston and the hills of West Virginia. By 1970, when VITA gained notoriety as a leading AT organization, its executive secretary would boast that it was the "chief source of technical assistance for all the 'hippy' communes in the USA."[21]

The history of VITA opens a tangled path through the politics of humane technology. Incorporated in 1960 with a high level of corporate engineering talent, the organization preceded widespread debates about authoritarian and democratic technics and complicates assumptions about the actors and vision behind appropriate technology. As historian Bess Williamson argues, "the small-scale technologies [volunteers] designed drew on an ideal of politically neutral technology transfer based on . . . practical issues rather than overarching ideological interpretations."[22] And yet, in the late 1960s, VITA was renowned, among development agencies, corporate executives, engineering educators, community action groups, and communes, not only for its humanitarian ambition but also because it claimed that the context of design—the "socio-economic factors, habits and customs that vary sharply"—was crucial to a harmonious technological society.[23]

The evolution of VITA from the scientists' movement to engineering mainstream, domestic and international AT agency, and finally enterprise institution, offers a microcosm of the intersecting energies embodied in engineers' search for humane technology. VITA's history reveals both the complications of a "non-ideological" solution to a professional crisis that was ideological at its core and how a savvy organization could thrive amid shifting professional ideals, funding structures, and visions of technology.

VITA got its start in the technological culture of Schenectady, New York, that inspired Vonnegut's tale of rogue engineers.[24] In 1958 the Mohawk Association of Scientists and Engineers (MASE) chapter of the Federation of American Scientists invited an official from the United Nations to discuss third-world technology transfer. GE research scientist Robert M. Walker was shocked to discover that the UN's technical assistance budget was smaller than the combined salaries of engineers and scientists in Schenectady. If the nation's technologists volunteered a few hours a week to individual problems, he calculated, they collectively would surpass all existing aid.[25]

Fed up with the failures of the antinuclear movement, a group of fourteen industrial scientists, engineers, and professors from GE, Union College, and the Knolls Atomic Power Laboratory created a Technical Assistance Committee under the auspices of MASE. In their simple plan engineers and scientists would foster peace and mutual understanding by solving problems through the mail. Aid workers and end-users in the developing world would send requests to a central dispatch, which would locate a relevant expert in a high-technology corporation or university, who in turn would answer from the comforts of home. Beginning with a single masonite chart, VITA developed a cross-indexed volunteer bank with thousands of entries that matched missionaries, Peace Corps volunteers, government agencies, and local businesses to technical professionals in projects that ranged from Nigerian soil chemists seeking techniques to rustproof typewriters to Maryknoll priests looking for battery-powered slide projectors or Indian entrepreneurs requesting product development expertise.[26]

The solar cooker project was instrumental to VITA's incorporation. In 1959 the humanitarian organization Cooperative for American Relief Everywhere contacted the MASE Technical Assistance Committee to solve problems of desertification in North Africa. Deforestation had severely reduced the wood supply, forcing women to dedicate hours searching for fuel to cook meals. Since the mid-1950s, development agencies had funded research on solar devices as an alternative to wood-burning stoves but failed to find a cost-effective device.[27] William Hillig, a GE physical chemist, hit upon a Fresnel mirror that could be built cheaply from Mylar and fiberboard. The design (which GE later credited to Johnson) resulted in a $25,000 seed grant from the US Department of Commerce, requiring the group to become a nonprofit agency independent of MASE. By 1961, when VITA brought its cooker to the UN International Conference on Wind, Solar and Geothermal Energy, it had a small office on the outskirts of downtown Schenectady with eighty volunteers.[28]

Volunteers initially encountered resistance from their employer on the grounds that they would compete with GE's international divisions and that they already benefited humanity through "regular work."[29] VITA, however, found an advocate in J. Herbert Hollomon, head of GE's General Engineering Laboratory, who gave the group managerial approval, made laboratory space available to volunteers, and encouraged the organization to appoint a paid director and advisory board of prominent technologists. Hollomon was so enthusiastic about VITA as an example of what engineers might achieve as societal leaders that, in 1960, he made it a centerpiece of his argument for a National Academy of Engineering.[30]

A grant from the Sloan Foundation generated the seed money that led to the hiring of Benjamin Coe, an MIT trained chemical engineer, as director and fundraiser. With assistance from Harvey Brooks of Harvard and Walker L. Cisler, president of the Engineers Joint Council, donations followed from the Rockefeller and Kettering Foundations, Detroit-Edison, GE, IBM, Standard Oil, Xerox, and others. Hollomon and Cisler's backing also helped VITA win a grant from the Engineering Foundation to develop partnerships with member societies such as the AIChE.[31]

In 1965 the *Reader's Digest* article "VITA has the Answer" vaulted the organization to a new scale. Stories of villages revolutionized by technical sketches received in the mail portrayed scientists and engineers as heroic servants and VITA as an organization that combined private enterprise with old-fashioned community values.[32] Inquiries jumped fourfold, and ballooned to 12,000 from 120 countries, which VITA tackled with volunteers from over 800 companies.[33] VITA also developed information services that included its lauded *Village Technology Handbook*, funded by the Agency for International Development (AID); and the Village Technology Center, a workshop for demonstrating small-scale technologies and training Peace Corps volunteers.

VITA's rapid growth had as much to do with engineering identity politics as the needs of the world's poor. While the organization initially solicited members through peace and social responsibility networks, it found an enthusiastic pool of volunteers by targeting the mainstream technical press. Early articles highlighted VITA's unique merger of humanitarianism, professional expertise, and low barriers to participation. A 1962 spotlight in *Product Engineering* that contrasted VITA's grassroots efforts with state-sponsored modernization tripled its membership.[34] Similar articles followed in journals from *Instrumentation Technology* to *RCA Engineer*.[35] As debates about out-of-control technology gathered steam, VITA's altered its pitch accordingly. "The engineer is basically a doer," its founders wrote in 1968, "like any other sensitive person . . . he is concerned about his fellow human beings." Playing to the search for authentic experience, they argued that VITA's "altruism and naturalness" offered "a meaningful outlet" to actualize their profession's better impulses.[36] This attitude helped spawn fifteen chapters across the United States, founded by charismatic individuals in local industries and universities that were only loosely connected with Schenectady.[37]

Though the majority of volunteers came from the corporate world, participation also was high among educators, most of whom saw engineering design as a tool of social and professional reform. From its headquarters at Union College, VITA boasted relationships with two hundred engineering schools. At Washington University in St. Louis, for example, VITA participation sparked nuclear engineer Robert P. Morgan to create a Department of Technology and Human Affairs.[38] VITA educators described the problems they tackled as an ideal confluence of "design" and "extreme ingenuity" that grew "out of a real cultural setting" and demanded a socio-technical approach.[39] The Catholic University of America, for example, created the I-Cube (International Innovations and Inventions) project, an "adventure in creative engineering" in which engineering students worked with peers from psychology, sociology, and economics. One team tackled the dual lack of public water taps and playgrounds in Peruvian slums by designing "play pumps," a solution described as at once aesthetic, simple, and human.[40]

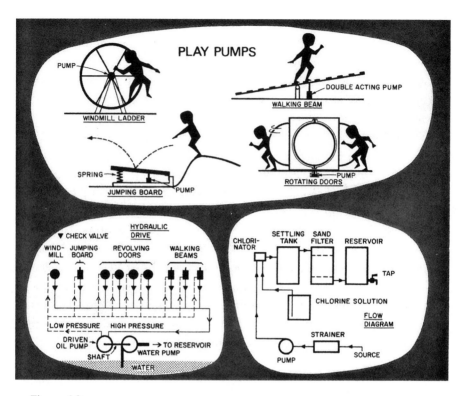

Figure 6.2
An exercise in creative engineering
Source: "Creative Engineering," *Industrial Water Engineering* 5, no. 8 (August 1968), 27.

In the late 1960s and early 1970s, VITA became a symbol of all manner of political visions under the banner of humane technology. *Civil Engineering* used VITA to illustrate engineers' "leadership potential as contributors to the solution of social problems," while *IEEE Spectrum* argued that VITA was a "socio-technical challenge" for the profession.[41] The Committee for Social Responsibility in Engineering printed a VITA recruitment pitch alongside an article on GE defense work.[42] Despite the fact that most of VITA's budget came from federal agencies, the *New York Times* stressed that members were "proud" their approach aligned with President Nixon's emphasis on "private voluntary effort and business involvement in anti-poverty and foreign-aid programs."[43] Meanwhile VITA received a glowing endorsement in the *Whole Earth Catalog*, which enhanced its visibility among the counterculture, leading to "almost daily requests" from domestic communes.[44]

For its part, VITA increasingly echoed themes of technological politics. While the organization maintained a focus on low-cost design and individual "change agents," it came to argue that achieving design simplicity was a complex systems problem that belied the assumption that "you can solve people's problems from your own basement workshop."[45] Board member Robert M. Goldhoff contended that VITA never imposed technology and stressed that users were treated as partners rather than academic subjects, while James D. Palmer, Union College's dean of science and engineering, asserted that "human dignity, hope and respect" were as important as technology to successful collaboration.[46]

VITA's evolving vision is best seen in its solar cooker project. Though it consulted an anthropologist early in the process, the cooker sprang from enthusiasm that expert invention could change the world.[47] In 1965, VITA earned an AID grant to test the cookers in Morocco. A UN agent convinced the governor of Fes Province to purchase 120 units, constructed by a local Peace Corps volunteer. When VITA checked in a year later, not a single cooker was in use. Reasons for abandonment ranged from unreliable heating to the fact that the governor's motive was to stop villagers from trespassing on his plantation. Citing the project as a failure on all fronts, VITA developed "broad principles" that stressed partners in the field, training for use and maintenance, and the primacy of cultural values in the design process.[48]

While VITA abandoned the cooker as a viable technology, it remained central to the organization's media narrative. GE deployed the cooker as an example of corporate technological progress in a socially conscious workplace. In *IEEE Spectrum*, cooker and engineer traveled from laboratory to field, where both stood between anxious villagers as a proxy for the engineer's moral reconciliation.[49] In a 1977 manifesto "AT: the Quiet Revolution," the *Bulletin of Atomic Scientists* used a decade-old photograph of the cooker as a representation of the libratory power of small-scale technology, with the engineer exiting the frame entirely.[50]

As VITA became a player in the politics of alternative technology, it found funding opportunities and ideological affinity for work in the United States. In 1969, it won $247,000 as the first in a series of grants from the OEO that doubled VITA's finances and partitioned it into international, domestic, and fund-raising divisions. By 1971, its budget reached a million dollars, with nearly 80 percent allotted for domestic work. The following year, VITA changed its name to Volunteers in Technical Assistance to reflect the trend.[51]

VITA-USA opened its first branch in Boston, where Route 128 engineers, professors, architects, and economists had been working on a combination of international and domestic assistance projects.[52] Subsequent chapters developed in Washington, DC; Houston; San Francisco; Colorado; and Philadelphia, which likewise attracted not just engineers and scientists but also lawyers, social workers, and accountants. Recruits tended to be from a younger generation concerned more with reimagining expertise than in bolstering it through acts of charity.[53]

Domestic work was fundamentally different than the international inquiry service. Projects ranged from investigating lead poisoning to organizing day care centers, collaborating with artists on modular playgrounds, constructing low-cost housing, and helping small businesses secure federal contracts. Work was immersive, often involving weeklong site visits that required company leave time. Projects also were more self-consciously political. Rather than an emphasis on heroic invention, VITA-USA stressed "soft technology" that actualized "underused, local *human* resources for solving community problems."[54] In Washington, DC, for example, VITA partnered with inner city gangs, while the free schools movement called the organization the most effective consulting service in the country.[55]

Bringing the small scale home, however, proved less tenable than VITA's international programs. For one, local work involved a face-to-face confrontation with the limits of technology as a tool for social change. Rather than semiprofessional intermediaries who shared their worldview, volunteers often engaged directly with politically marginalized users. Moreover, though industry initially expressed enthusiasm about VITA's domestic growth, funding never materialized because companies wanted their own names attached to local aid and because they were cautious about the critique inherent in VITA's approach to domestic poverty. Most significant, the organization failed to match federal funds with private donations at the same time that the Nixon administration broke up the OEO, distributing its functions to an array of agencies. The bottom fell out in 1972, when OEO funding was not renewed, forcing the termination of half of VITA-USA's staff. VITA-USA limped on with regional funding until 1976, having worked on nearly thirty-five hundred projects.[56]

VITA-International encountered its own challenges against the backdrop of VITA-USA's rise and fall. The high watermark in requests came in 1969, and that of volunteers in 1970, a pattern shaped by a decline in federal funding for international development.[57] Moreover, as VITA entered its second decade, it was becoming more of a professional NGO than a volunteer organization. Director Beno Sternlicht, a former nuclear engineer at GE, implemented a system that used phone inquiries and pamphlets developed by paid staff to solve problems more effectively than volunteer correspondence. An external review further pushed VITA to value efficiency over participation, while lauding its "decentralized" and "adaptive process" as a model of innovation for development institutions.[58] Then, in 1973, VITA moved from Schenectady to Washington, DC, where its paid staff grew to forty employees.

Amid internal upheaval, VITA became a central player in the AT movement. Led by Thomas H. Fox, a former Peace Corps official, VITA collaborated with Schumacher's Intermediate Technology Development Group; was a founding member of Private Agencies Cooperating Together, a network of technological, scientific, and agricultural aid; contributed to both the National Center for Appropriate Technology in Butte, Montana, and AT International in Washington, DC; and published nearly forty thousand how-to manuals a year.[59]

In a 1978 congressional hearing on AT, Fox presented VITA's interpretation of AT, asserting that the "determining factor of appropriateness" in a project was the user rather than "technology itself." He described an ideal solution as one that was "labor-intensive" versus "labor-displacing," based on local expertise and resources, locally sustainable, "requested by and defined by the local consumers' and users' participation," sensitive to the local environment and values, and that, most important, achieved actual benefits to "low-income people."[60]

Nonetheless, AT was controversial both in VITA and the agencies and corporations that funded it. Many VITA founders argued that cutting-edge scientific research was needed to make a real difference. In 1969, for example, Walker pushed for utilizing satellites to assess cropland viability.[61] Others located the most fruitful sites of change in small business and industry. In his congressional testimony, Fox navigated a middle path. Warning against the "easy panacea" or "political credo," he lamented the partisan dynamic of AT. Charges of "frivolous gadgetry" or that "big must be ugly," he argued, stemmed from a failure to have a "serious national and international dialogue about technology and technology choice." Influenced by VITA's own tumultuous experience, Fox argued that a socio-technical reformation would come with the merging of domestic and international efforts and a rejection of the dichotomy between high and low tech. Above all, he concluded, AT was "a way of thinking and of placing values on technology . . . [that] mandates that the choice of technology should result from a *process* . . . and the process, once set in motion, has its own potential for promoting change and development."[62]

As VITA navigated the shifting tides of development funding in the late 1970s, its approach evolved toward entrepreneurship in an interpretation of AT as low–capital-intensive growth. It developed manuals on implementing small-scale technology for the multinational firms Cargill, TRW, and John Deere. And, in the early 1980s, after winning a $10 million grant from AID to develop an international network of alternative energy resources, VITA established a Venture Division and abandoned the rhetoric of AT.[63] The organization would continue to thrive as an innovative development agency, merging in 2005 with the NGO *Enterprise Works*. VITA, however, had long ceased to be a movement for the engineering profession to change itself as it changed the world.

Twentieth Century da Vincis

In April 1969 an assemblage of discarded motors, bicycle wheels, and miscellany rescued from a New Jersey landfill graced the cover of *IEEE Spectrum*. The smoking wreck, captured in the process of bashing itself to bits, opened a two-part series on the history of avant-garde art that implored engineers to take up yet another creative revolution.[64] Readers learned that the kinetic sculpture was one of the earliest projects of Bell Laboratories employee Billy Klüver, founder of Experiments in Art and Technology, an organization dedicated to shaping technology that *Chemical Engineering* called "not the preconception of the engineer or the artist but rather the result of the human interaction between their two areas."[65]

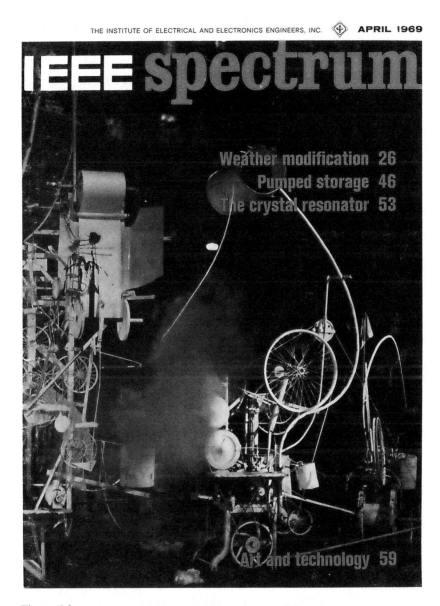

Figure 6.3
Homage to New York in *IEEE Spectrum*
Source: *IEEE Spectrum* 6, no. 4 (April 1969): cover. Courtesy of the IEEE.

Largely on account of such profiles, E.A.T. was an instant phenomenon among engineers, one that bore striking parallels to VITA. The technical press argued that the turn to art would help engineers "find brand new goals."[66] The art world swooned that E.A.T. "was essential in a crumbling democracy."[67] Leading theorists of technology exchanged notes with Klüver on the meaning of his enterprise. And, over two thousand engineers, scientists, and technicians joined E.A.T. in twenty-six chapters from New York to Japan with hopes of making technology human.

E.A.T.'s projects have achieved canonical status as exemplars of the nascent technological arts. Beginning in the late 1950s, artists and musicians shifted from abstract expressionism to assemblage, performance, and chance events. By the mid-1960s they were exploring cybernetic feedback, holograms, lasers, and video, and increasingly doing so through brokered corporate partnerships. E.A.T. was not alone in this milieu, but along with MIT's Center for Advanced Visual Studies (CAVS), it served as a leading access point to advanced technology and publicity for new media artists.[68]

Approaching E.A.T. from the perspective of the social history of engineering reveals a different dimension of encounters between the two cultures in the 1960s. For technical professionals art was not a neutral or incidental pursuit. It conjured contradictory values of hybridity and purity; elite expertise and participatory democracy; the neutrality of knowledge and its inherent politics. As engineers struggled for recognition in science's shadow, art resonated with the visual, object-oriented dimensions of design. In *An Introduction to the Art of Engineering,* for example, Alvin S. Weinstein and Stanley W. Angrist of Carnegie Mellon defined the engineer as "an artist who begins with an idea or need" and then uses his "special tools," constrained by time and budget, to bridge the "creativity gap" between theory and reality.[69] At the same time corporations and research laboratories appealed to the arts as a way to convey institutional soul. The Airborne Instruments Laboratory of Cutler Hammer, for example, showcased employee artwork to prove that its staff was "civic-minded or artistic and generally a broad-gauged group of humans" of the sort "the world needs."[70] At MIT, engineering students flocked to art courses affiliated with CAVS. The main commonality in this evolving landscape was the certainty that a merger of art and technology was a source of intellectual, professional, and, indeed, moral good.

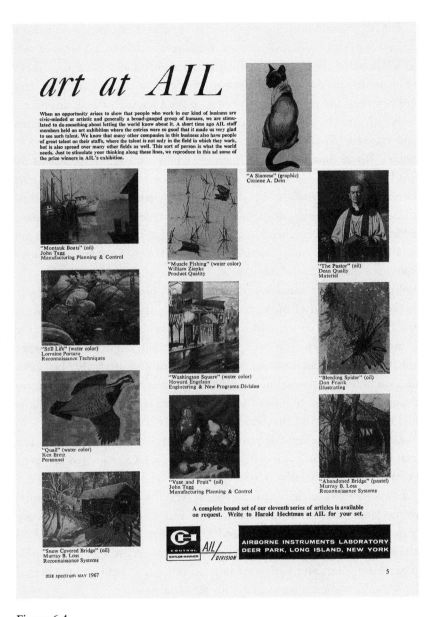

Figure 6.4
Art at AIL
Source: Cutler Hammer, "Art at AIL," *IEEE Spectrum* 4, no. 5 (May 1967): 5.

E.A.T. channeled this diverse set of aesthetic commitments into a formal but loosely defined network. Engineers sought to aid artists much as VITA aided the developing world, but also to battle their own sense of alienation, to demonstrate the humanistic value of their vocation, and ultimately to use technology to bring about a society of new meaning.

Billy Klüver, E.A.T.'s iconoclastic founder, was born in 1927 in Monaco and raised in Sweden, where he received a degree in physics at Stockholm's Royal Institute of Technology while studying film at nearby Stockholm University. He then moved to Paris for a job at Compagnie Générale Thomson-Houston, GE's corporate partner, where he continued to cultivate art-world connections. In 1954, he left for Berkeley to earn a PhD in electrical engineering. After a brief teaching stint, he took a job at the Communication Science Division of Bell Laboratories, forty minutes southwest of Manhattan.[71]

In 1960, Klüver aided his friend the Swiss-born artist Jean Tinguely in the construction of *Homage to New York*, the machine featured on *IEEE Spectrum's* cover, which self-immolated at a happening at the Museum of Modern Art.[72] Collaborations soon followed with Jasper Johns, Robert Rauschenberg, and Andy Warhol. Klüver also began to fashion himself as a cultural critic, arguing in the art-house review *Hasty Papers* that "system builders" had drifted from societal service and thus had to reconnect with the "individual realities" of users.[73]

Driven by the belief that artists were society's pioneers and seers, but that they had become isolated from technology, Klüver invited scores of artists to tour Bell Labs. These encounters lay the groundwork for a major collaboration. In October 1966, at New York's 25th Street Armory, twenty engineers, ten avant-garde artists, and ten thousand visitors contributed to *9 Evenings: Theatre and Engineering*. The event included an electronically coordinated tennis match, chance recordings by the composer John Cage, and a maze of inflated tubes that fed light and sound back as the audience interacted with the structure.

Despite malfunctions and negative reviews, the crew regrouped after *9 Evenings* to announce the formation of E.A.T. At a meeting of over three hundred artists, engineers, and journalists, Bell Laboratories researcher Fred Waldhauer outlined the rationale for the new group by explaining that engineers believed art was as "important as science and technology" to society and that artists believed "sensitivity to technological media" was essential if the arts hoped to remain relevant to human experience.[74] To bridge the gap, E.A.T. would initiate a series of lectures for artists on cutting-edge technologies and develop a technical services program to match artists with engineers for individual collaborations.

E.A.T.'s founding partners, Klüver and Rauschenberg, saw their project as an "organic social revolution" in which altering the design process would alter the character of the technological society.[75] To harness technology's humanizing potential, responsibility needed to be taken not only by engineers and artists, but also by corporations, which Klüver argued could extend their innovative capacity by providing artists access to their prohibitively expensive tools.[76] The association would benefit all parties. Artists would realize their vision, engineers would learn to "do different things—look at things differently," and companies would reap the ideas and patents.[77]

E.A.T.'s normative vision varied, however, as Klüver sought bridges with a range of institutions and movements. He first looked to the engineering establishment in an unsuccessful attempt to gain approval from the IEEE for an "Engineering in Art" section. He encountered similar ambivalence at Bell Labs, which declined his request to recreate *9 Evenings* as a pedagogical exercise at MIT.[78] He finally found a partner when Theodore W. Kheel, a prominent labor leader, offered E.A.T. space in the American Foundation on Automation and Employment's "Automation House," a nineteenth-century Manhattan brownstone wired with cutting-edge technology. Kheel, the former head of the National War Labor Board, believed that collaboration between labor and management could resolve automation's unintended effects.[79] He was strongly influenced by the Harvard Program's interpretation of technological change, and, in a twelve-page insert in the *New York Times*, linked Automation House and E.A.T. to "A philosophy for living in a world of change." Technology, he wrote, "can overwhelm us if we let it. But it offers rich rewards for human development if we are willing to confront and control it."[80]

Klüver's own philosophy was a loose amalgam of concepts drawn from opposite poles of the technology & society genre, which he modulated depending on his audience. Klüver was dismissive of professional critics—including Snow and McLuhan—but nonetheless deployed their rhetoric in pursuit of his interests. When he resigned from Bell Labs in 1968 to dedicate his life to E.A.T., he cited Galbraith's *New Industrial*

State to accuse engineers of neglecting the consequences of their labor. He compared E.A.T. to VITA and the antinuclear Pugwash as a means of "mobilizing scientists and engineers" to social purpose.[81] But, in a letter to Mesthene, he declared in the same breath that technology had "no inherent values" while championing its nature as a "cohesive not a disruptive force."[82] Klüver was equally unclear about E.A.T.'s role as a social movement, at times arguing that E.A.T. was "a political organization" and at others contending that it was strictly "nonpolitical."[83]

E.A.T.'s open-tent approach to collaboration reinforced the organization's ideological ambiguity. On the one hand, it attracted hundreds of engineers and artists because it stressed the social conditions of design in a call for a participatory democratic process that resonated with theorists of technological politics. On the other hand, there was a strong undercurrent of faith that technology's inevitable thrust of change was breaking down old hierarchies, and that engineers and artists needed to adjust themselves to new social realities if they wanted fulfilling lives.

E.A.T. promoted itself to engineers as a democracy with no aesthetic talent required. Recruits were enticed via publicity in trade journals. Others signed up when the electronics entrepreneur Seymour Schweber donated his company's booth at the 1968 IEEE Convention. Still others joined after the juried competition *Some More Beginnings*, where prizes were awarded to the most creative technical contribution in a collaboration. Many were intrigued by *E.A.T. News*, which described aiding artists as a "challenging context" requiring "professional skills" in a "human situation" that was "real and exciting . . . meaningful."[84] Finally, chapters grew locally in far-flung corporate networks. At Ampex, for example, a cafeteria lunch discussion became a group of a few hundred art-minded engineers.[85]

Participants in E.A.T. included students, professors, corporate employees, NASA research scientists, and DOD cold warriors. They attributed their interest to artist spouses or told of "dabbl[ing] in oil painting." For every traditionalist, there was an engineer who composed electronic music or built cybernetic art. At least half claimed no artistic connection: "Just a curious Midwestern engineer," wrote one electronics specialist from Ohio.[86] In addition to information about their skills and availability, engineers often sent E.A.T. lengthy statements of existential purpose.

Bruce Steinberg of Berkeley's Space Sciences Laboratory, who would go on to become a famed album cover designer for the likes of Jefferson Airplane, described himself as "the funkiest continuously-employed draft-deferred field engineer in the Apollo program." He explained how an upbringing with a "technician/teacher father and artist/musician mother" led to his "hybrid" status in a world of "two cultures," and how E.A.T. could help him overcome a "terminally disillusioned" corporate life.[87]

Though E.A.T. facilitated over five hundred matchings between artists and engineers, successful collaborations were the exception rather than the rule. Artists were frustrated by engineers who worked sporadically during leisure hours. Engineers, for their part, complained that rather than aesthetic partners, they felt "used," as little more than "a paid engineer, cold, without feeling."[88] In other cases, projects sparked ideological clashes. For example, one artist's plan to build a jet pack roiled a Douglas Advanced Research Laboratory engineer who argued that personal flying machines, which had "existed in our collective storehouse of fantasy for centuries," made a mockery of the need for "imaginative alternatives" in a "milieu that threatens to destroy human life and values."[89] Another, however, saw the project as "an exciting challenge," offering "many happy liftoffs."[90]

With little direction from E.A.T. beyond an initial contact, it was unsurprising that collaborations were stilted; revolutionary social relations were not expected to come easily. Lack of corporate interest proved a more vexing problem. E.A.T. tried to win over industry by claiming that collaborations generated marketable spin-offs, the first of which purportedly occurred during *9 Evenings*.[91] But IBM, Xerox, and other companies made only token donations of $1,000 to receive recognition as official sponsors. So E.A.T. relied instead on Kheel, a few entrepreneurs, small foundation grants, and members' dues for its livelihood. Even Bell Labs was skeptical. At Bell Labs, engineers and artists engaged in open-ended projects with new media, including electronic music and video. Manfred R. Schroeder, for example, investigated the information content of visual imagery, and his colleague Ken Knowlton, another computer graphics pioneer, applied these techniques to one of E.A.T.'s most recognized works, a computer plot of a nude, compiled from mathematical symbols.[92] When projects were in the capacity of research, there was little friction. When they took place publicly and cut into engineers' time, company willingness was grudging. Division director John R. Pierce—himself a science fiction writer and creativity theorist—argued that E.A.T. was "strictly comparable to golf, skiing, politics, public service, and other spare-time avocations."[93]

In fact every attempt to obtain sustained corporate support for E.A.T. failed—until its largest project, the Pepsi Pavilion. In 1968 a Pepsi executive responsible for the company's entry at the 1970 World Expo in Osaka, Japan, decided he would do something "genuinely prestigious."[94] He lived next door to Robert Breer, an animator, sculptor, and former engineering student. After convincing Breer that Pepsi would support avant-garde experimentation, Breer in turn convinced E.A.T. that it would be given full control over the project.

The Pavilion was an impressive technical feat. Designed to embody "choice, responsibility, freedom, and participation," it had no set path for visitors and no crowd control such that visitors could stay for hours or stroll through in a few minutes.[95] E.A.T. planned a rotating crew of artists who would direct the environment in accord with their individual vision. Outside, the artist Fujiko Nakaya and the cloud physicist Thomas R. Mee cloaked the structure in a "natural" fog.[96] Inside, E.A.T. assembled an inflated spherical Mylar mirror and built an audiovisual system with a krypton laser.[97] The final cost was over a million dollars, twice the agreed-upon budget.

Frustrated by E.A.T.'s freewheeling style, Pepsi enraged the crew by requesting that E.A.T. replace its collage of recorded sounds with a familiar Japanese nursery rhyme.[98] E.A.T. refused, and asked Pepsi to legally recognize the Pavilion as a work of fine art.[99] Just before the Expo's opening, Pepsi instructed E.A.T. to cease and desist. The crew scrambled to remove its programming tapes. Lawyers later reconciled rights and expenses. But E.A.T. returned to the United States a shell of its former aspirations.

When collaborations oriented to the fine arts proved no longer viable, E.A.T. attempted to retool as a community action organization. It built a prototype greenhouse for urban environments on the Automation House's roof; conducted an experiment in which children could play with fax machines, telephones, and closed-circuit televisions in an unstructured learning environment; and developed an ill-conceived plan to temporarily convert a cubic mile of Los Angeles into recreational space.

Figure 6.5
Pepsi Pavilion Source: Photo Courtesy of Shunk-Kender © Roy Lichtenstein
Foundation.

E.A.T.'s efforts to expand beyond art, however, received almost no
funding in an already crowded nonprofit field.

E.A.T.'s problems were exacerbated by a sea change in attitudes about
technology among artists that exposed political tensions at the center of
art and engineering collaborations. Throughout E.A.T.'s existence, the
arts frequently had served as politics by other means. A few months after
E.A.T.'s *IEEE Spectrum* profile, for example, former IEEE staff writer and
Burndy History of Science Library assistant director Gordon D. Friedlander offered a counterplea for "planning, order, and logic" in his own
millennial history of aesthetics. Rejecting "electronic fads and shock
effects," he argued that real collaboration would occur when the "truly
creative engineer" worked with classically trained artists, as they had
in antiquity and the Renaissance.[100] But while the virtue of creativity
remained universal, if disputed, collaboration took on sinister connotations. Factions in an art world that once lauded technological collaborations as the future of aesthetics attacked E.A.T.'s corporate premise as
aesthetically bankrupt and as a veiled cover to deflect attention from military R&D. One underground newspaper went so far as to declare E.A.T.
a front for the CIA, responsible for stealing the creative ideas of artists.[101]

At the same time the Pavilion fiasco brought Klüver face to face with hard truths about changing either art or engineering. As his dream unraveled, he lamented the era's "cultural stagnation" and criticized the moralism of technology & society debates, which, he argued, crippled actual projects.[102] Most of all he resented the co-optation and dismissal of his vision's revolutionary character. By 1971, E.A.T. was in financial ruins, forcing Klüver to sell much of his personal art collection. In December 1972, amid frayed personal relationships, Klüver and his remaining followers were kicked out of the Automation House.

Despite its abrupt collapse, E.A.T. was far from a failure. Though the new media vision E.A.T. promoted experienced a backlash in the early 1970s, it has developed into one of the most dynamic areas of technology development in venues like MIT's Media Lab. Engineers in these contemporary boundary spanning aesthetic environments likewise view their work as linked to a revolutionary technological future in which they can express themselves creatively and humanely without becoming entrenched in politics. They might ponder the fact that, in the late 1960s, where debates about the military-industrial complex left lasting fissures, artistic collaborations at worst were labeled a distraction to "real" engineering.

The Change Managers

One of the hallmarks of the technology & society genre, reprinted in anthologies and debated in college classrooms, was Paul Goodman's pithily titled essay "Can Technology Be Humane?" Beginning with an account of MIT's March 4 work stoppage, the anarchist philosopher compared the nation's crisis to the Protestant Reformation. According to Goodman, "the entire relationship of science, technology, and social needs both in men's minds and in fact" needed to be remade in a "responsible learned profession" that embodied "prudence," "modesty," "ecology," "decentralization," and "creativity."[103]

Published in the *New York Review of Books* in November 1969, "Can Technology Be Humane?" is remembered as a measured response to the neo-Marxist critic John McDermott's scathing *NYRB* review of the Harvard Program, "Technology the Opiate of the Intellectuals."[104] But Goodman's original audience was not the academic Left. His essay first appeared four months prior under the more ominous title "The Case Against Technology," nestled between articles on "Marketing High Technology at Honeywell" and "How a Technical Man Can Invest." Its readers were a self-described "set of men very important to this country" known as the Innovation Group.[105]

The Innovation Group announced itself with a dire warning. Full-page advertisements in a range of publications from *Scientific American* to the *Wall Street Journal* explained that modern industry was on the verge of self-destruction for failing to control technology. Student radicals were entering the corporate world demanding solutions. All the while, the vertically integrated corporation was dissolving as a result of its inventions. For a fee of seventy-five dollars, members would receive the Group's journal *Innovation* and join a "project of discovery" to define "the skills and the standards of a newly meaningful profession."[106]

Augmented by telephone-based salons, electronic databases, and executive workshops, the Innovation Group was more than a clever gimmick for selling magazines. *Innovation* was to spark a "process of interaction" in which managerial problems were shared, government policy debated, and entrepreneur/venture capitalist connections made. The Group's staff would be "facilitators" and "environmental creators" of a multifaceted professional who partnered with social critics, architects, community organizers, and religious leaders to fight bureaucracy with creativity, launch entrepreneurial ventures, and save society from runaway technology.[107]

The Innovation Group's high-powered advisory board included OTA architect and former congressman Emilio Daddario, Harvard Program director Emmanuel Mesthene, and J. Herbert Hollomon, the one-time GE research director who had protected VITA in its infancy. Among its featured guests were former Secretary of the Interior Stewart Udall and the economist Milton Friedman. By 1971, the Group's ranks climbed toward ten thousand.[108] Then, in June 1972, as abruptly as the Group appeared, it unceremoniously vanished.

To the extent that scholars have recognized the Innovation Group at all, it has been far less heralded than VITA or E.A.T. Yet it contributed to the most lasting conceptual legacy of engineers' collaborative turn. Rather than defining humane technology through outreach, the Innovation Group sought to become the creative vanguard within the corporation

How Can You Deal with the Forces of the 70s?

New pressures are making the task of today's manager more difficult than ever. Standard management practices are often ineffective. Routine solutions don't remedy new and complex problems.

At the same time, new brush fires spring up, even as old ones are spreading. As a result, management is driven to "reacting" to situations — rather than planning for them. Short-term projects take precedence over long-range objectives.

"Crisis management" describes these self-defeating practices whose folly is apparent even as we implement them. But in the face of uncertainty, how do you avoid crisis management? How can you, as a manager, devise a strategy that helps your organization capitalize on the forces of change rather than suffer them?

First you have to identify the forces and trace their complicated interrelationships. Only then can you begin to develop the plans for action that will ensure your organization's survival — and your own as well — amid the intensified changes of the 70s.

That's why **The Innovation Group** has built its third annual Conference/Workshop around the theme **"The New Forces: Management's Challenge and Response."** Its purpose is to help you gain new insights and expand your perceptions of the nature of the decade ahead and how it affects your business future. For three days, October 3-6, you will have the opportunity to exchange views with distinguished members of the international management community, from Herman Kahn to Koji Kobayashi.

You will participate as a member of a workshop team. You will engage in roundtable discussions . . . confer with panelists in informal sessions . . . view closed-circuit broadcasts . . . and add your own comments via direct phone lines.

Acting as catalysts for the free-flowing exchange of ideas are the panelists, who represent a broad range of industries and functions, and the program, which explores a diversity of business problems. (See card opposite.)

If you have felt the impact of the new forces, you owe it to yourself to take advantage of the opportunity that ⒸⓌⓈ affords to enlarge your perspective.

LAST CHANCE TO REGISTER — USE THE CARD OR WRITE

Carol Conrad, Conference Coordinator
The Innovation Group
265 Madison Avenue
New York, N.Y. 10016

ⒸⓌⓈ **October 3-6, 1971   Harrison House   Glen Cove, New York**

Figure 6.6
The Innovation Group
Source: Innovation Group "How Can You Deal with the Forces of the 1970s," *Innovation* 23 (August 1971): 64.

itself. Simultaneously terrified and awed by the social movements challenging their authority, the Group's members asserted that the corporation was both generator and victim of accelerating change and that there were parallels between Route 128 and Palo Alto spin-off firms and communal and new Left factions. Running throughout was the belief that once technology's fatalists and utopians recognized that "innovation" was the only path in an era that defied total control, problems of conformity, access to tools, and personal growth could be resolved by entrepreneurial leaders.

At a first approximation, the Innovation Group was the result of a 1968 partnership between William G. Maass, a charismatic publishing executive; Robert Colborn, an engineer turned magazine editor; and Jack A. Morton, the Bell Laboratories research director responsible for bringing the transistor to market. But the Group's mercurial existence belies a more robust network that was a decade in the making. The Group channeled a confluence of factors including federal projects to extend the lessons of the microelectronics industry to the civilian economy, professional strategies to distinguish engineering from science, corporate public relations campaigns to blunt technology's critics, and management theorists' efforts to gain organizational insight from the very same critics.

A rationale for the establishment of a National Academy of Engineering distinct from the NAS was the belief that, on the one hand, pure science couldn't capitalize on new knowledge, and, on the other, that the free market lacked means to serve the public good. At the same time, in the upper echelons of federal science policy and the social circles of chemical and construction industry managers, there was extensive hand-wringing about the scale of military and aerospace funding.

In 1961 Jerome Wiesner and John Kenneth Galbraith asked President John F. Kennedy to create a development commission to stimulate civilian industry. The Department of Commerce was already charged with this mission, but it was a relatively weak agency with few R&D functions. To spearhead new government-industry partnerships in the model of postwar military sucesses, Wiesner chose Hollomon to serve as Commerce's assistant secretary of science and technology.[109]

Hollomon would have met a casting call for the engineering profession's postwar ideal. The graduate of a Virginia military prep school, he earned his PhD in metallurgy from MIT during World War II. He then ascended swiftly to GE's highest research position, where he was responsible for betting on the market potential of scientific advances. He also maintained an adjunct position at RPI, served as an advisor to Cornell and MIT, and was a founder of the NAE.[110]

Hollomon saw the engineer as the catalyst for humanitarian capitalism. In 1960, before innovation was a commonplace term, he argued that the process of integrating scientific, organizational, and social knowledge into actionable change—what he called "planned innovation"—heralded "a way of life as revolutionary as the concept of the importance and dignity of man that wrought the Renaissance, and as forceful as the concept of capital formation that initiated the industrial revolution."[111] A study he chaired for the EJC demonstrated that postwar breakthroughs were the consequence of a 77-fold increase in federal R&D funding. Despite the successes, 90 percent of resources went to just three agencies (the Atomic Energy Commission, DOD, and NASA) and accounted for nearly 85 percent of aerospace R&D. By contrast, just 2 percent of the domestic construction industry's anemic R&D budget came from the government. In an argument that presaged Melman's *Pentagon Capitalism*, Hollomon claimed that civilian industries consequently were stagnant and had started to lag Japan, Europe, and the Soviet Union. The root of the problem was not just budgetary; it was a failure of professional self-definition. Engineers emphasized devices rather than social systems, and worshiped science at the expense of "the needs of people and society."[112]

Hollomon found support for mobilizing government and industry in an NSF study on "patterns and problems of innovation." The report's lead author Donald Schön, a young jack-of-all-trades at Arthur D. Little, Inc., proposed a theory of "innovation by invasion" to explain how mature industries changed only when challenged externally; for example, the chemical industry's impact on textiles with the invention of Nylon. But, Schön argued, the electronics industry presented a "strikingly different" pattern. In one of the first analyses of the microelectronics revolution, he described a research-based industry built form scratch. A single organization—Bell Laboratories—fostered an interdependent system of universities, government laboratories, small businesses, and multinational corporations in which specialized products were translated into mass markets with stunning rapidity.[113] How, Schön and Hollomon asked, could this model be emulated in other sectors while mitigating the disruptive effects of innovation?

In 1962 Hollomon presented Congress with a project to bring planned innovation to traditional industry. His Civilian Industrial Technology Program (CITP) would be a partnership between private businesses and experts at a new agency in the Department of Commerce. The CITP would operate a technical services program—similar to the decentralized approach that later defined VITA and E.A.T.—to foster regional university/business centers on college campuses, award R&D contracts, and attract scientists and engineers to regions outside the aerospace and electronics boom. The CITP, however, was a lesson in technocratic failure. Aiming for a $50 million program, Hollomon received less than $1 million. He encountered resistance from budget hawks and existing agencies protecting their turf. But the CITP ultimately was hobbled by the construction industry it purported to help, which viewed Hollomon's plan as federalism run amok.[114]

Hollomon continued to pursue government-industry partnerships as the source of humane technology, turning his attention to the character of the engineer. He helped make creativity, innovation, and entrepreneurship a defining element of the NAE's policy vision. In 1966 Commerce, the NAE, and the NSF co-sponsored a workshop in which nearly a hundred like-minded technical leaders championed the "creative processes of invention and innovation," which Hollomon argued, were the source of "socially and politically feasible" solutions to the nation's socio-technical problems. According to Hollomon, entrepreneurship was a form of "creative engineering" that could be taught but required a design-centered curriculum of scientific training mixed with economics and social knowledge.[115]

Hollomon's vision fell squarely within the purview of an ideology of technological change. As he championed the engineer-entrepreneur, he shared conference stages and NAE planning sessions with Mesthene, Simon Ramo, and like-minded thinkers.[116] But Hollomon saw a deeper democratic realignment than the redirection of aerospace firms to urban development. In a keynote panel at the 1969 IEEE Convention, he argued for "basic and fundamental changes" in engineers' attitudes toward society. Citing McLuhan's notion that technology had made an "'all-at-once' world," he excoriated his profession to rousing applause. "This is an 'us-ness' world. It is a world of community," Hollomon contended. "Yet in fact there is very little sense of community. Your technology has made it almost impossible for a man to be himself." Technological change presented opportunities for "intimacy" and "humanity." Instead, Hollomon lamented, advances in electronics and communications had been used to "centralize."[117]

Frustrated with his failure to foster innovation inside the federal government and opposed to President Johnson's escalation of the Vietnam War, Hollomon resigned in search of new challenges. Optimistic that academe could give postindustrial capitalism its moral compass, he became president of the University of Oklahoma and began to remake its lines of authority through a collaborative process involving participants from all levels of the organization.

At the same time that Hollomon, Wiesner, and others were building their case in Washington, a group of entrepreneurial journalists joined the cause. In 1960 William G. Maass, a corporate vice president at the industrial trade publisher Conover-Mast, had his pulse on the emerging concept of innovation. According to Maass, the "consumption of fundamental science by technology" was breaking down identities, occupations, and nations.[118] In a booming market for public understanding of science, he saw an unmet need for a different kind of magazine that would give meaning to rapid change for scientists and engineers themselves.

Maass spared no expense in realizing his project, *International Science and Technology* (*IST*). His editorial team consisted of journalists who had crossed boundaries between academe and industry, expert and public, technical and social. Senior editor, Robert Colborn, for example, was a Dartmouth civil engineering graduate and published novelist who had served as managing editor of *Business Week*, while executive editor Daniel I. Cooper was a nuclear physicist, with a PhD from MIT, who left a job at Bell Labs to become managing editor of *Nucleonics*. Funded by corporate advertising, *IST* was distributed free of charge to 120,000 scientists, engineers, and research managers, which Maass characterized as the top 10 percent of the world's "responsible decision-making technical men."[119]

IST carried traditional features such as interviews with Nobel Prize–winning physicists, but it found the true drivers of progress in coverage of "The University and Regional Prosperity" and "The Science Entrepreneur."[120] Feature articles and interviews by Wiesner, Hollomon, Schön, and others did not simply echo socio-technology ideals, they cultivated them. The 1964 article "From Research to Technology," written by Jack A. Morton highlights *IST's* role as a conceptual incubator.[121] Working with staff editor David Allison, Morton developed his first published thoughts on the innovation process. Morton walked readers through Bell's organizational flowcharts, historical documents, scientific principles, and company authorizations to produce a generalized theory of technological change. He followed up in 1966 with a report on the "Microelectronics Dilemma" that stressed the need for a theory of the "people-process of innovation" in order for companies like AT&T to "survive and grow in a new era of technology."[122]

In 1968 Maass and his team moved from describing the new era to shepherding it into being. When *IST* found itself in the middle of a merger between two giants of the industrial trade press, Maass and Colborn paired with Morton to create Technology Communication, Inc., the parent company of the Innovation Group. Building on *IST* connections, its founders developed a core set from elite technology managers and establishment intellectuals.

The Innovation Group's fourteen-member advisory board of high-technology managers was active as authors, keynote speakers, and builders of the Group's network.[123] They accounted for close to half of *Innovation's* output. Board member Warren Bennis, for example, the president of the University of Cincinnati, argued that the student movement was driven by a "person-centered" morality that sought "joy, spontaneity, self-expression and self-actualization." Youth were amenable to work in high-tech industries, but "only if the technology serves personal growth and social goals."[124] Mesthene likewise paired with Hollomon to describe the need for corporations to become socially responsible. According to Hollomon, in a networked society, managers needed to "participate cooperatively" with other industry leaders and the government in public markets to improve social services like education and health care.[125]

The Group developed experimental information networks to extend their message. Describing *Innovation* as "the first magazine with an interactive feedback system," it built an "electronic meeting place" that facilitated over seven hundred reader/author conference calls.[126] Contributors heralded a future of "degree-granting Woodstock festivals" for technologists to learn on a drop-in, drop-out basis, which the Group subsequently pioneered in a seminar on "Explorations in the Management of Technology."[127] Conducted at a Long Island estate with a registration fee of $350 (equivalent to $2,000 today), the workshop included theater talks by Hollomon, Morton, and famous entrepreneurs. Random generated small-group discussions and live transmissions of lectures to hotel room televisions with interactive question and answer sessions were intended to foster creative collaborations.[128]

The Innovation Group embodied tensions between those who saw themselves as apostles of a socio-technical revolution and those who saw business as lacking an articulate response to technology's critics. Some contributors were true believers in a countercultural reformation. Former Lockheed engineer turned psychologist John M. Steele Jr., for example, paired with *Innovation* staffer Ron Neswald to argue that: "an Aquarian lifestyle is not necessarily incompatible with the sophisticated technocracy."[129] Reader response, however, found that even Goodman's sympathetic critique met with hostility or indifference.[130]

Much of *Innovation*'s tempered coverage of political radicalism was designed to identify new markets and diffuse political tension. In an editorial titled "The Sophisticated Spokesman," for example, *Innovation* laid out a doctrine for dealing with confrontational tactics, particularly for organizations involved in military R&D. Among the suggested steps was the creation of company journals that reprinted articles about "technological fallout" to satisfy employees and learn the language of the critics.[131] In this respect one of *Innovation*'s major functions was to advocate a set of positions that bordered on sloganeering.

But the Innovation Group should not be dismissed as a mere generator of management buzzwords or as a front for "the system." *Innovation* provided a forum in which leading organization theorists developed and communicated their research. The clearest example can be found in Schön's contributions to the Group. A professional nomad from his earliest training as a jazz pianist to his PhD in Deweyian philosophy, Schön joined Hollomon at the Department of Commerce shortly after penning his NSF report on innovation. He then left to form his own nonprofit social research company, the Organization for Social and Technological Innovation. In 1967, on the strength of his book *Technology and Change: the New Heraclitus*, Schön was appointed to the faculty of MIT's Sloan School.

Schön articulated the most nuanced explanation of the engineering profession's malaise. In a pragmatist philosophy that echoed Goodman, he contended that one did not have to look far to see that technological change had undercut its own promises. A great part of contemporary turmoil, he argued, was a Baconian obsession with the "Technological Program" in which technology and society could be controlled. The accelerating pace of change, however, had shattered the myth of a "stable state" and along with it hopes of a progression "toward a good society whose objectives remain stable and clear throughout." Among its most pervasive effects was a loss of professional identity.[132] For all its pitfalls, technological change was the state of nature and needed to be approached with a flexible mind. Schön argued that "diffusion" systems of organization were being replaced by "learning" systems in military-industrial projects, social welfare programs, and, most of all, radical social movements, which he described as "a loosely connected, shifting and evolving whole" fueled by "an ethos in which transformation around the new is a value in itself."[133] Rejecting the Movement's philosophy of revolt, Schön nevertheless championed its "ethic of change" as a model for professional life. To thrive in a networked society, individuals and organizations required constant experiment and self-reflection, living in "here-and-now" and seeking "the new."[134]

For all of its advice on surviving a revolution, however, the Innovation Group lasted just three years. Its demise was shaped by tragedy. In September 1970, Colborn died of cancer. Then, in 1971, Morton was found murdered in the charred remains of his car, the victim of a mugging gone wrong.[135] The following year, *Innovation* was quietly absorbed by the liberal economics journal *Business and Society Review.*

When the Group disbanded, its editorial staff moved to corporate public relations, media consulting, and freelance journalism where they continued to promote change management. Its board members also found themselves in positions that mirrored their writing. When Harvard pulled the plug on its Program on Technology & Society, Mesthene left to become dean of a Rutgers University experimental college for nontraditional students. Hollomon left Oklahoma for MIT as special advisor to president Wiesner and director of the Center for Policy Alternatives (CPA), a new unit that was in many respects a successor to the Harvard Program, but with an engineer at the helm. At the CPA, Hollomon led a comprehensive overview of engineering education as well as a mapping of the nation's technological future based on eight different policy directions. He insisted that understanding "the nature of technological change itself; i.e., the *process* of change" and how policy choices impacted technology were the most vital tasks confronting the nation's leaders.[136] As he had argued throughout the decade, scientists and engineers offered the most creative opportunities for reform and government participation with industry was vital to confront technology's secondary effects.

Conclusion

In a world of iPads, Facebook, and the global proliferation of smart phones, "humane technology" appears as a self-evident material and discursive reality. In the pages of *Wired* magazine, the pixels of BoingBoing, and the stages of TED, there is no shortage of praise for the design revolution of the late 1960s and early 1970s. In these forums Brand, Fuller, and McLuhan are honored prophets for identifying technology as a source of inevitable democratic liberation.

It is tempting to interpret engineers' collaborative reforms as part and parcel of this restoration. At the dawn of the 1970s visions of humane technology reached well beyond E.A.T., VITA, and the Innovation Group into virtually every domain of the profession. At Kaiser Aluminum & Chemical Corporation, for example, employees received a series of technicolor newsletters that blended psychedelic imagery and quotations from McLuhan, Fuller, the I Ching, and Alan Watts to describe a societal shift from "structure to process" taking place on the streets, in the classroom, and in the R&D office.[137] Process-oriented design also appeared to be the future of engineering education. The 1974 textbook *Design: Serving the Needs of Man* (formerly titled *Elementary Problem Solving*) opened with a litany of crises from famine to urban sprawl. Its many illustrations included images from E.A.T., and its margins reprinted over two hundred aphorisms from Archimedes, Mumford, McLuhan, President Nixon, and engineers in the Innovation Group.[138]

But as the trials and tribulations of engineering's collaborative partnerships show, "humane technology" was infused with adversarial qualities. The central tension was the relationship between the meaning of *technology* and the meaning of the *engineer*. Virtues of creativity and collaboration provided the impetus for new social and textual networks, but they also proved the space for conflict. Engineers in VITA, E.A.T., and the Innovation Group navigated between technology's critics, counter-cultural champions of the design revolution, and the conservative institutions and norms of their profession. In these networks their interpretations of "humane technology" were bound up in historic images of the engineer and broader efforts to make remake the material world in the context of ideologies of technological politics and technological change. As reformers in confrontational political movements had discovered, this confluence of ideology and identity politics was not easily surmounted.

The conversion of Samuel Florman from the engineering profession's reconciler to its chief partisan drives home the extent to which cultural politics shaped engineers' visions of humane technology. While a practicing engineer, Florman wrote a Master's thesis on Kafka at Columbia University. To share his passion for the humanities, in 1968, Florman compiled a five-year cultural enrichment program for working engineers that introduced them to Snow, McLuhan, and Mumford, before turning to Western classics.[139] In his course reader, he claimed that engineers experienced oneness with the world similar to artists. Strengthening the similitude between the two, he argued, would make engineers "wise, sensitive, humane, and responsible."[140] But as humane technology collaborations ran their course, Florman grew disillusioned. "The anti-technologist has no prior claim on creativity," he told an ASCE design conference in 1972, "Quite the contrary. If anyone has a prior claim it is the technologist."[141] A decade later, his spear had sharpened. Making no mention of VITA, he excoriated the AT movement as "small is dubious."[142] Reviewing the *Next Whole Earth Catalog* for the *New York Times*, he rejected its politics: "There is . . . no sign of commonwealth— no industry, no government, no museums, no universities—nothing but the credo of the self-absorbed individual." From his perspective, instead of a historically informed vision of prudence and professionalism, humane technology had evolved into an "unrealistic faith in primitive technologies and a growing mood of selfishness masquerading as self-reliance." Still he held on to the dream of Proteus, arguing that: "*mutual reliance* is what civilization is all about."[143]

In the decades that followed, the personal computer industry helped restore technology's public image along lines championed by the Innovation Group, while engineers never fully overcame charges that they were "dull," "uptight," "lonely," "inauthentic," "half-men."[144] To understand the re-enchantment of technology but the failure of engineers to reinvent themselves as its humane creators, we need turn at last to the nation's engineering schools.

7

Making Socio-Technologists

"As inevitable as night after day"—using the word "inevitable," contrive six similar figures of speech. "As inevitable as . . ."
—George C. Beakley and H. W. Leach, *Engineering: An Introduction to a Creative Profession*[1]

In a secluded house on the outskirts of Boston, a group of twenty-eight insurgents plotted. Their gathering began with self-pity over failures to alter the status quo, but after four days of freewheeling discussion about drugs, music, and anarchy, they regained their nerve. "On Monday we go to Los Angeles and take over the goddam division . . . form a cell of two people, find an ally. . . . Gather additional members." Political revolution and bassoon-playing electrical engineers on LSD are not typical subjects of engineering pedagogy workshops, but these were the themes that occupied educators and student activists from the nation's leading technical schools at MIT's Endicott House in June 1968. At the center of the conversation was the American Society for Engineering Education survey *Liberal Learning for the Engineer*, presented to the group by the study's director Sterling P. Olmsted. "I'm convinced that we live in an age of major revolution in science, society, and technology," Olmsted pleaded, but the relevant stakeholders "wanted to go on with their own little jobs."[2] His outburst was prompted by the group's handwringing over the recent polemic "Stamp Out Engineering Schools," in which Robert Hutchins of the Center for the Study of Democratic Institutions called for the abolition of MIT, Georgia Tech, and Caltech.[3] The group later heard from Caltech's student president, Joseph Rhodes Jr., who organized an interdisciplinary air pollution study with eighty other students. The meeting concluded with a commitment to initiate similar projects before the nation's youth gave up on engineering careers.

In the previous chapters, I explored how engineers in professional member societies and a range of alternative organizations appropriated social theories of technology to redefine notions of responsibility, service, and what it meant to be human. Most of these endeavors had a foothold in the academy. Academe's centrality to both the cold war economy and its dissidents meant that universities were among the first sites to give voice to competing theories of out-of-control technology.

It was as institutions of formal training, however, that the university served as the most significant venue for instilling technology's meaning to engineers. Engaging students not yet initiated into engineering culture with alternative visions of technology was a means of normalizing what counted as relevant knowledge. Professional leaders moreover considered student attitudes about technology to be one of the largest uncertainties in engineering's continued growth. Indeed the Endicott participants returned to their institutions to find Dow and GE recruiters assailed by protesters and graduate students no longer exempt from the draft.[4] In 1970, the ASEE, whose "humanistic social" division they had fantasized taking over, issued a directive admonishing faculty to keep engineering classes running during campus strikes.[5] By 1972, the nation's freshman engineering enrollments were down 35 percent from their 1967 peak.[6]

Reformers, however, faced an uphill battle in their desire to integrate social inquiry about technology in the making of new engineers. Universities are both notoriously resistant to change and almost continuously involved in "crisis" and "reform." Whatever vision to which they aspired, reformers had to navigate bureaucracy, financial constraints, local traditions, and the students they wished to mold.[7] More to the point, champions of a socio-technical perspective worked in the shadow of one of the most successful reform movements in the history of American engineering education.

Pursuit of an engineering science ideal had made engineering schools models of the "federal grant university."[8] In the mid-1950s, land grant colleges, which previously operated on limited external funding, were receiving millions of dollars for engineering research. Meanwhile such traditional staples as draftsmanship and surveying were curtailed or eliminated altogether as classroom instruction turned toward advanced mathematics and computer programming.[9] This transformation had been built with the cooperation of government, industry, and professional organizations.[10] By the mid-1960s, a generation of faculty had been cultivated and tenured in this mode and hundreds of thousands of students were taught in its methods. Reformers argued that the curricula of the cold war era exacerbated the perennial problems of undervaluing social knowledge in engineering education and carried with it a "hidden curriculum" that stifled creativity and autonomous thought.[11]

The background of the Endicott conference pointed to an additional complication in shaping new engineers. Its attendees were employed in engineering schools and self-identified as engineering educators, but their training was in English, history, anthropology, and political science. What's more, they frequently were antagonistic to the professional mainstream represented by the ASEE, which they saw as hopelessly conservative.[12] Often this antagonism was mutual. Integrating social knowledge with technical practice, in other words, presented reformers with unique challenges that circuits or thermodynamics did not. On the one hand, such knowledge was beyond most engineers' formal competency and typically was taught by professors who engineers claimed knew nothing about engineering and were hostile to its purposes. On the other hand, if reform-minded engineers could direct socio-technical pedagogy or reach common ground with humanist and social scientific collaborators, they could make it serve the profession's goals.

Challenges aside, among engineering's visionaries, there was near unanimous consent that some kind of socio-technical pedagogy would enhance the profession's reputation, solve society's environmental problems, and reverse the trend of America's unraveling.

Culture Wars in the Classroom

To educators with historical perspective, calls to make engineers more humane had a familiar ring. For almost as long as engineering careers in the United States have begun in the academy, reformers have targeted the humanities and social sciences to reinvent who engineers should be. In the happier days of the postwar boom, educators advocated training engineer-citizens in the service of cold war democracy. In 1956 the ASEE conducted a detailed survey of the "humanistic–social stem" of professional education that echoed James Bryant Conant's *General Education in a Free Society*. Its report reiterated the aspiration of earlier studies that 20 percent of a student's coursework come from humanistic and social subjects. Course summaries outlined how professional leadership was to be cultivated by studying Shakespeare, John Steinbeck, Plato, John Locke, and the history of technology.[13] These humanistic-social courses nonetheless were considered supplemental to the training of engineering scientists.

In Sputnik's aftermath, however, few stakeholders were content with the state of engineering education. In 1961 the Engineers' Council for Professional Development requested that the ASEE conduct yet another national review. The six-year study, *Goals of Engineering Education*, emphasized the challenges of engineering science. It portrayed the profession as caught in a bind. Scientific knowledge was expanding while its useful lifetime contracted. New problems necessitated integrating scientific knowledge with creative design.[14] The *Goals'* major and most controversial suggestion was to designate the master's degree as the "first professional degree." But it also stressed that humanistic-social training could set the "genuine engineer" apart from technicians and scientists as a man who saw the "system of the future as a whole."[15] In short, humanistic-social studies might give engineering enhanced credibility with respect to science and nation.

As charges against technology mounted, existing humanistic-social courses seemed less than adequate. Educators hosted workshops that brought together engineers, humanists, and social scientists to master the myriad new texts in the technology & society genre, but these interdisciplinary conversations did little to quell the volume and tenor of critique. Meanwhile the escalating disorder of the nation's campuses only exacerbated the sense of urgency among engineering educators.

It was in this tenor of crisis that Olmsted, a Quaker literary scholar, conducted the ASEE's 1968 survey *Liberal Learning for the Engineer*. Faculty teams visited 27 colleges and queried 179 others. Olmsted found that 93 schools had initiated programs in liberal studies since 1965.[16] He highlighted exemplary courses that differed radically from those of 1956. General education gave way to student choice, and Western civilization was jettisoned for action-oriented seminars. Canonical texts in the western tradition remained, but Mesthene, Mumford, and technology

& society anthologies now dominated syllabi. But his report characterized most reform as superficial rather than as a rethinking of engineering
in the "total human culture."[17] It offered no definitive models but stressed
the importance of developmental and contextual goals rather than utili-
tarian skills or content coverage and found hope in curricula co-designed
by humanists and engineers.

By 1973, goaded by Olmsted's report, nearly two hundred of the
nation's technical colleges experimented with curricula to address
technology's social implications.[18] These efforts exhibited consider-
able heterogeneity even as the ASEE and the ECPD tried to impose
common standards. Educators implemented combinations of four broad
approaches: (1) "humanizing" engineering through interdisciplinary
liberal education, (2) teaching systems analysis to produce professional
socio-technologists, (3) introducing nonengineers to technological thinking, and (4) creating social-scientific experts outside of engineering.

Humanistic engineering programs flourished at elite universities and
a small group of liberal arts colleges whose faculty were drawn to theories of technological politics. At Dartmouth, for example, the course
"Crisis of Human Values in a Technological Society" explored:

Value judgments implicit or explicit in the decisions made by engineers, scientists,
and planners in attempting to deal with such problems as local and global pollution, quality of the environment and of life in general, military funding of
research, secret work, world food and population problems, nuclear deterrence,
genetic control, and the man–machine relationship.[19]

At Tulane, the Kent State shootings prompted a mechanical engineer and
an English professor to collaborate in designing a new course. "Since
some of the power of high technology over the values of man results
from the mystique derived from its myth and its complexity," its instructors explained, engineers needed to reduce technology's "autonomy and
. . . power to assist in the objectification of man through understanding
its nature and the way it works."[20]

Design engineers also embraced the notion of "humane" engineering,
which resonated with techniques for selecting alternatives and spotlighted
the aesthetic and social qualities of entrepreneurship. Paul B. Daitch at
RPI, for instance, cited Mumford, Commoner, and Wiener to argue that
engineering education needed to focus on "why" in addition to "how."
"Design," he asserted, was "the major vehicle which relates technique

Figure 7.1
Aesthetic scapegoat
"This student's mechanical man flapped its arms, burst into flames and destroyed itself." Source: Gale E. Nevill Jr. and John A. O'Connor, "Unbottle Your Creative Ideas: A Cooperative Venture in Engineering and Art," *Engineering Education* 63, no. 2 (November 1972), 116. Courtesy of Gale E. Nevill Jr.

and society."[21] To emphasize creative problem-solving in this vein, one University of Florida class, jointly taught by an artist and an engineer, constructed Tinguely-like self-destroying machines and "scapegoat[s]," aesthetic objects designed to "relieve hostilities and release aggressive feelings."[22]

A more common approach within engineering schools—including state universities and polytechnic institutes—sought to create experts in the mold of Ramo's socio-technologist. These programs received rhetorical support from member societies and industry, which preferred a less fundamental critique of technology's human values. In 1970, for example, the NAE workshop "Social Directions for Technology" endorsed the notion of a "uniquely endowed cadre of *social engineers*."[23] In a guest editorial in *Engineering Education,* the president of Gulf Oil likewise called for socio-technical training. Quoting Mesthene, he argued that it had been nobody's obligation to attend to technology's unintended effects, but a new kind of engineer could do so within private industry.[24]

Plans to implement such curricula ranged from individual coursework to major foundation sponsored efforts at UCLA, Case Institute, and IIT.

Most pedagogical reform was targeted at undergraduates, but technology & society experiments also were incorporated into graduate training. David P. Billington and Robert Mark at Princeton, for example, argued that a historical perspective in advanced technical training allowed civil engineers to better analyze their designs. In 1968 they developed a graduate seminar on aesthetics for a "new type of engineer who . . . will center his career on a union of technology with the humanities."[25] At RPI, the economist Edwin J. Holstein and systems engineer A. Bruce Carlson invited social theorists, including Mesthene and Thayer Scudder of Caltech, to lecture in their systems theory course "Engineering in Its Social Context."[26]

Staunch defenders of the profession, however, rejected the notion that engineers needed to be more humane. Instead of trying to reform engineers, they sought to instill technological thinking in the general university population. At the University of Alabama, James H. Black created the course "Engineering: A Cornerstone of Our Society" to defend engineers as open-ended problem solvers and assert that pollution was a consequence of giving "society what it wants."[27] Similarly Lafayette College's Center for Engineering Programs for Non-Engineers tracked eighty colleges with programs to reduce biases among the "technically illiterate . . . unprepared to cope with or even to grasp" the contemporary world.[28]

Finally, academic social scientists situated in technical schools claimed that they, rather than engineers, should be primarily responsible for directing technology to social purpose. While the Harvard Program was the most heralded example, the trend was notable also at Caltech and MIT where economists and political scientists built programs of international reputation. Courses in the economics of technological change, risk management, and urban policy proliferated. These centers for sociotechnical research developed somewhat apart from engineering faculty. By the late 1960s, however, social scientists, engineers, and natural scientists reached across campus in exchanges that laid the groundwork for graduate programs in Science, Technology, and Society (STS).

Despite a sometimes tenuous relationship between engineering and liberal arts faculties, the role of humanists and social scientists stood out in efforts to reinvent engineers. Throughout the postwar era, literary scholars and a new breed of technology historian—including Thomas P. Hughes, Melvin Kranzberg, Carroll Pursell, John Rae, and Lynn White Jr.—acted as advisors and instructors for engineering students and faculty. They assimilated technology & society literature to engineers' sensibilities and in many cases were ambivalent about technology's critics.[29]

This affinity for technology as a positive social force shaped how humanist and social scientific educators guided engineering students through texts. Kranzberg and Pursell's 1967 anthology *Technology and Western Civilization* began in the Stone Age to demonstrate technology's centrality to humanity. Developed at the request of the United Armed Forces Institute and Extension branch of the University of Wisconsin, the book hinted at technology's "ambivalence," but concluded that the "acceleration of technology" made the twentieth century the "American Century."[30] Later anthologies and monographs written for—and sometimes by—engineers moved technology's critics from the endnotes to the front page.[31] Noel de Nevers, a chemical engineer at the University of Utah, for example, published the annotated collection *Technology and Society*, created for his NSF sponsored classes. De Nevers apologized for the gloomy tone of his book, but felt he had an obligation to confront the "Pollyannas" who refused to believe the "problems are real."[32] A later book of the same title by Richard C. Dorf at University of California, Davis, began simply: "Technology takes over." Critical of the "technological imperative," its first chapter alone reprinted excerpts from Marine, Goodman, Illich, and Galbraith. Dorf instructed that technology could not solve every problem, but that liberally educated engineers might be able to better understand and control its forces.[33]

General engineering textbooks aimed at high school students and college freshman likewise strove to inculcate socio-technical vision. Existing texts revised their introductions and added design exercises to bolster themes of social responsibility and creativity. The 1969 edition of *Engineering as a Career*, for example, supplemented aerospace case studies with a VITA exercise on the irrigation of a Filipino rice field that reprinted correspondence between an engineering professor and a Peace Corps worker in the field, as well as homework assignments that included finger painting.[34] The most ambitious effort to shape public attitudes about engineering was the series of books, laboratory exercises, and visual aids known as *The Man Made World: A Course on the Theories and Techniques that Contribute to our Technological Civilization.* Funded by NSF, Bell Labs, and IBM, it was conceived as "a part of the cultural curriculum—a course for all citizens who will take part in guiding the currents of our society."[35] Its directors E. E. David Jr. of Bell Laboratories and John G. Truxal of the Polytechnic Institute of Brooklyn were both science advisors to President Nixon. They were assisted by forty-six contributors from engineering colleges, high schools, Bell Labs, and IBM. The project began in 1964 as a general introduction to engineering, but as it reached fruition Truxal described it as a response to the "nationwide anti-technology sentiment and to the young people's accusation that engineering education is irrelevant."[36] Its underlying vision was of a world whose survival hinged on the "ability to adapt to these technological developments and to control the changes so that they remain in the realm within which society can adjust."[37] The 1973 edition used the example of the medical pacemaker to illustrate six lessons: once a technology was introduced, "the public in general tends to take it for granted"; the pacemaker was a spin-off from the "space and military programs" attacked by critics; it was "difficult to anticipate and control the side effects"; the public was "insatiable" and demanded constant improvement; the pressures of new technology extended into nontechnical domains, complicating the engineers' task; and regulatory control and national standards became essential because no individual could monitor technology's multiple impacts.[38]

Surveys, workshops, and textbooks are useful for locating social networks and common knowledge, but desire for reform met institutional reality and actual students on the local level. Reform moreover was driven as much by internal motivations as national ones. To bring into relief the challenges the engineering academy faced as a site of normative guidance for the profession, I want to focus on four schools whose pedagogical transformations are especially illuminating. First, I examine the major peaks in what Clark Kerr hailed as the "California mountain range"—UCLA, Caltech, and Harvey Mudd College (HMC).[39] These institutions represent a comprehensive state university, an elite private research institute, and a unique experiment in liberal engineering. Finally, I turn to MIT, the nation's preeminent example of cold war engineering and its discontents.

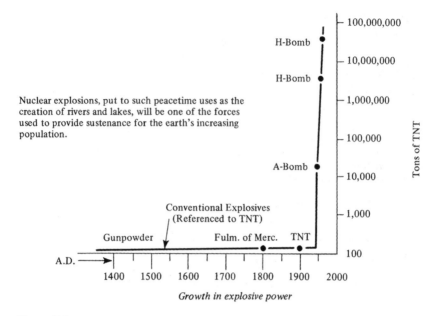

Figure 7.2
Charting change
Source: Engineering Concepts Curriculum Project, *The Man Made World: A Course on the Theories and Techniques That Contribute to Our Technological Civilization*, vol. 1 (New York: McGraw-Hill, 1969), 23.

Optimizing Human Values at UCLA

The fact that UCLA did not even have a College of Engineering until 1945 speaks to the extent of California's educational ascendency during the cold war. Founded in the waning days of World War II, UCLA's College owed its existence to southern California's aerospace boom, in-migration, and a baby boom that tripled California's population between 1940 and 1970, transforming UCLA from a commuter school into a central node in the multiversity.

Professor Llewellyn M. K. Boelter was sent from Berkeley to head the new College. He brought with him a neoclassical vision in which education was meant to impart technical skill with moral content.[40] Privileging undergraduate education, he created a unified curriculum in a single department to train professional leaders rather than specialists. Students received a degree in general engineering with freshman and sophomore years devoted to physical sciences, mathematics, and graphics; followed by advanced training in materials, mechanics, circuit analysis, engineering design, thermodynamics, and engineering economics.

But California's demand for technical manpower and UCLA's desire for federal research funding pulled it in new directions. By 1955 the original faculty of seven had grown to eighty-six. Undergraduate instruction expanded from twenty offerings in 1945 to eighty-five in 1955. Initially the College offered only two graduate courses, by 1955 there were twenty-four.[41] The College's development prompted it to reexamine its purposes, and in 1957 it won a $1.2 million grant from the Ford Foundation to undertake systematic curricular review.[42] Extending over twelve years and resulting in forty-one reports and fifteen textbooks, UCLA's Educational Development Program (EDP) was the largest site evaluation of engineering pedagogy in the postwar era. Its "most startling discovery" was not that engineers needed to emulate scientists, but rather that "*expertise* in the applied humanities [was] both the outstanding *need* of the professions and the most exciting challenge of professional education."[43]

The EDP's motive force was Allen B. Rosenstein, a magnetic-systems engineer who had been one of the first recipients of an engineering doctorate from UCLA. Rosenstein stressed that "design" was the "essence of engineering."[44] For Rosenstein curricular planning was itself a systems problem subject to mathematical analysis.[45] Quoting Julian Huxley, the EDP team contrast science, where "values are deliberately excluded," to engineering, which was a process of "making the future" that required an "operational value system."[46]

A subcommittee led by engineers Bonham Campbell and Jacob Frankel worked with William H. Davenport, a professor of English at HMC, to make humanistic-social studies an integral part of professional training. As it stood, such coursework made up less than 20 percent of the curriculum, with minimal integrated elements. In 1963 the committee argued that a required humanities course needed to be taught within the College of Engineering.[47]

Materials engineer Daniel Rosenthal described the EDP's curricular changes as a Hegelian synthesis of past reform—report writing and its antithesis of cultural finishing would be superseded by applied humanities to make engineers "more creative and human."[48] To that end, in 1966, he and Davenport co-designed the upper division course "Engineering: Its

Role and Function in Human Society," intended for both engineering and humanities students. It confronted technology's critics asking: "Are engineers on trial?" "Are humanists on trial?" And "for what?" Its reader, which became a bestselling textbook, reprinted excerpts by authors as varied as Herbert Hoover and Ellul. It concluded with the category "hope for the future" so that engineering and liberal arts students could see their collaborative path to a progressive future.[49]

As debates about technology escalated, Rosenstein came to view the EDP as a means of making a new kind of professional. He believed that humanists had abandoned relevance, leaving engineers responsible for "generating new values for the society of his design objects."[50] In the EDP's final report—*A Study of a Profession and Professional Education*—Rosenstein argued that society was in the midst of a "massive" societal crisis. If steeped in the design of optimal human value systems, engineers could combine scientific and humanistic knowledge to alleviate its ills.[51]

Desires aside, the EDP's goals were at variance with UCLA's direction. Chauncey Starr, president of the Atomic International Division of Rockwell, succeeded Boelter as dean in 1966, and implemented swift changes. Starr liberally employed the rhetoric of the humanities and social sciences in support of cost–benefit analysis to confront technology's ill effects, asserting that engineers needed to take "a lead responsibility for the guidance of our social development."[52] Nonetheless, at the very moment that an integrated humanistic-social engineering education came to dominate the national scene, it was discontinued at UCLA in a further scaling up of the engineering sciences. The College became the School of Engineering and Applied Science to stress graduate training, divisions were formed by specialization, and freshman and sophomore courses were eliminated in favor of a pre-engineering curriculum.[53]

After the failure of his vision, Rosenstein worked to develop a National Professions Foundation (NPF), analogous to the NSF, that would help engineers become society's leaders. He described the NPF as a natural outcome of EDP studies and argued that the NSF's basic research orientation hindered engagement with technology's ill effects, while the NPF would provide a "quantum jump . . . in social responsibility."[54] In a decade-long campaign, he recruited engineering educators, top managers at GE and GM, and colleagues in law, medicine, business administration, and urban planning. In 1980 his proposal met its unsuccessful end on the floor of the US House of Representatives.[55]

Though its plans faltered at UCLA, the EDP was taken as a promising model by engineering educators nationally. It exhausted its 3,500-copy supply of *A Study of a Profession*, which Hollomon of MIT's Center for Policy Alternatives described in 1973 as the only logical means of curricular reform. IIT moreover used UCLA's reports as a blueprint for its own lauded reforms.[56] Still, Rosenstein's unfulfilled plan is instructive of reform's challenges at any institution competing for grants, personnel, and students in a system where success was measured in terms of quantifiable technical virtuosity.

Caltech's Aims and Goals

A decidedly different sort of engineering institution lay just a short drive inland from UCLA. By the 1960s the words "engineer" and "engineering" rarely appeared in Caltech publicity, and 60 percent of its students majored in physics or math. Where UCLA's student body had grown rapidly, Caltech's remained stable and small. Its faculty expanded in the cold war era, but not nearly at UCLA's pace. Caltech's size, prestige, and homogeneity helped it sustain a mission of fundamental research. However, by the mid-1960s faculty worried about student unrest and the backlash against technology. Drug use and attrition were rising as students bemoaned Caltech's alienating culture.[57]

In seeking solutions, "humanizing" Caltech became entangled with making social-scientific experts. Among the largest differences between education at Caltech when compared to universities like UCLA was that humanists and social scientists were an institutional minority. Its Humanities Division began as a non–degree-granting service division devoted to teaching future scientists and engineers. For most of Caltech's history, students took a required core in literature, US and European history, economics, and public affairs.[58] In the 1960s, the role of the Humanities Division changed. Caltech eliminated its humanities core in favor of electives and introduced a BS in economics, English, and history, intended to produce scientifically literate graduates for political leadership. The biggest change, however, was the expansion and composition of its faculty. MIT's high-profile economics and political science programs were a source of envy that fueled a series of new research oriented hires, including RAND economist Burton H. Klein. In 1967, the Division was renamed Humanities and Social Sciences to describe its expanded mission.

The social scientists set the tone of Caltech's engagement in debates about out-of-control technology, fostering quantitative analyzes of technological change. Also in 1967, Caltech embarked upon an $85 million development campaign under the banner "Science for Mankind." President Lee DuBridge explained that science always had served mankind, but in the past it had done so "haphazardly." Maximizing science and technology's "good and helpful impacts" and minimizing "adverse and painful ones" now required direction through interdisciplinary research in the social and behavioral sciences, humanities, and environmental engineering.[59]

Students, however, remained frustrated with the gap between Caltech's rhetoric and its actual reorientation to social action. Student president Joseph Rhodes Jr. argued that Caltech was "producing highly skilled technicians" rather than well-rounded, responsible professionals.[60] He wanted to reinvigorate scientific and engineering progressivism through "alternatives to traditional methods of structuring education and organizing research," and helped a team of students win a grant from the National Air Pollution Control Administration that involved technical research on smog and social experiments with "sensitivity labs," T-groups, and student interaction with corporate consultants.[61] His success as a tempered reformer earned him an invitation to MIT's national meeting of engineering education at Endicott House and then a position to President Nixon's Commission on Campus Unrest, and ultimately sparked a political career as a three-term US Congressman.

In 1967, technology & society debates extended to the administrative level when dissent in a faculty meeting prompted an Institute-wide review of Caltech's societal obligations. For the next two years a Committee on Aims and Goals explored plans to reform the Institute. Proposals included converting the Jet Propulsion Laboratory to an environmental science laboratory, admitting women, having aerospace majors study the politics of the supersonic transport, and creating a new college of arts and humanities. The faculty expressed reservations about this challenge to put Caltech in mankind's service. They feared the Institute would drift from its mission and dilute its excellence.[62] Attempts to attract first-rank humanities students would fail. Federal funding was declining and new programs might divert resources from science and engineering divisions. Finally, conservative faculty members rejected the idea of "relevancy," arguing that the pursuit of knowledge for its own sake was the essence of intellectual inquiry.

The social sciences offered a contentious but mediating position. Frederick Thompson, dually appointed as a computer scientist and philosopher, argued that Caltech should avoid action-oriented studies or educating the "culturally deprived." Instead, it could investigate the underlying theory of "increasing rates of technological and social change."[63] Caltech's social scientists asserted that they could provide the Institute's normative guidance. Science, they claimed, "recoils from the wished-for or the what-ought-to-be" and technology was not "good or bad in a social or moral sense," thus its values depended on "whom the engineer hires himself to serve." Economics and political science could address "use-judgments," because those fields had become the "applied sciences" of greatest relevance, but achieving success required a graduate program.[64]

The debate over integrating social knowledge into engineering education thus became a debate *between* the humanities and the social sciences. Social scientists saw Caltech as uniquely poised to create expert policymakers for a technological society. Critics warned against propagating scientism, and defended the humanities as a source of judgment that provided "not only a counterbalance but a valuable complement in the Caltech educational process."[65] While pedagogical philosophies were at stake, scholarship was the dominant factor in decisions about staff, direction, and curriculum. The Division that emerged made concrete the tension between human values and instrumental reason.

Engineering as Liberal Learning at Harvey Mudd College

In 1957, as the nation's engineering schools labored to retrofit their curricula to the engineering sciences, the first new engineering college in three decades opened its doors with liberal learning as its mission. Comparing itself to experimental colleges of the 1930s, a third of HMC's curriculum would be devoted to humanities and social science so that its students could "assume technical responsibility with an understanding of the relation of technology to the rest of society."[66] A strong base in science and mathematics was essential to modern engineering, according to president Joseph B. Platt and the college's trustees, but HMC did not want to produce applied scientists. The humanities and social sciences would foster "intellectual penetration" "analytical ability" and "values."

At the start, William H. Davenport, former head of English at the University of Southern California, chaired a humanities faculty of two that included him and an assistant professor of English. Like general education elsewhere, HMC's curriculum consisted of expository writing, literature, and western civilization with electives available in Claremont's Associated Colleges. And, with John Rae's arrival from a tenured position at MIT, in 1959, HMC added the history of technology.

But HMC was an early innovator of cross-disciplinary and interdisciplinary coursework. In 1960, the systems engineer Warren E. Wilson replaced freshman graphics with a project-oriented design course. An open-ended problem was chosen yearly by engineering faculty in consultation with humanists and social scientists. Students were divided into teams that were required to consider nontechnical factors in their solutions. In 1961, HMC also introduced *Science and Man's Goals* a problem-oriented senior seminar, co-taught by a humanist and an engineer. In the fall students examined man's individual relationship to science and technology, and in the spring, the impact of technology on society.

By 1966, HMC attained operational stability. The faculty reached critical mass, the campus took shape, students enrolled in an expanding array of courses, and graduates were either accepted to graduate school or gainfully employed.[67] Faculty and students, however, wondered if HMC was meeting its goals. Attrition was high, students were grade-driven, and social knowledge seemed to have fallen into a secondary role.

HMC's crisis of confidence came to be framed as a struggle for the core of its mission. There was broad consensus for further integrating the humanities with the technical curriculum. Proposed reforms included creating nontraditional majors to train students capable of "managing, coordinating, interpreting, and criticizing technology"; teaching technology & society courses that would bring together engineering and liberal arts students from other Claremont Colleges; and overhauling the entire curriculum.[68]

Initial reforms mirrored the evolution of Davenport's evolving philosophy of technology. In 1968–1969, he spent a sabbatical at the Harvard Program on Technology and Society, where he wrote the *One Culture*, a monograph that revisited Snow's divide with the premise that technological change was humanity's great challenge and that the only way to master it was for humanists, sociologists, and engineers to collaborate.[69] He modeled a new required freshman course at HMC, *Man, Science, and Society*, on the Harvard Program's orientation.[70]

With a grant from the Sloan Foundation, HMC planned even deeper changes. In 1970, it implemented a common freshman year divided into *Mathematics*; *Natural Philosophy* (physics and chemistry); and *Quest for Commonwealth*, a program "concerned with fundamental problems that modern man encounters both as an individual and as a member of a society." *Quest* was the vision of Theodore Waldman, who had spent a two-year leave of absence at Berkeley's experimental Tussman College. It presented a different direction than *Man, Science, and Society*, which hewed closer to the Mesthene and Ramo ideals. *Quest's* goal was to introduce students to "the sources and nature of humane values" and to explore their relevance to technology. Themes included "war and peace," and "freedom and authority." All freshmen read the same texts, discussing them in an assembly of the entire HMC community and then in small seminars.

Quest's goal was to instill that "understanding the political legal and social problems that men had dealt with was as important as suggesting a solution to them." Students read great books from Thucydides's *History of the Peloponnesian War* to contemporary social theory. A central theme of this longue durée approach was that some problems lay beyond scientific or technical solutions. Waldman described *Quest's* ideal exercise to be a model of "a legislative body deliberating over a public issue" in which competing values were ever-present.[71]

On the heels of its newfound energy, HMC introduced a revised mission statement that served it into the 1980s. It stressed the similitude between sonnet and equation, the human basis of all technical problems, and the cultivation of humility. "Insist that tools take you only so far," it admonished, "then apply what you know to making a better life for yourself—a better world for others."[72] *Quest for Commonwealth*, however, was discontinued in 1974 as overtaxed faculty returned to research, the imperatives of tenure, and the comforts of disciplinary identity.

Technology Studies at MIT

An undergraduate of MIT's class of 1972 had myriad opportunities to learn the social meaning of technology while earning credit toward a degree. He could participate in an urban transportation laboratory, enroll in an experimental college with a self-paced curriculum, investigate technology policy with former congressman Emilio Q. Daddario, take a "champion gut course" with former SDS president Carl Oglesby, or pursue a hybrid major in engineering and social inquiry with neo-Marxist critic John McDermott.[73] One might not fault him if he was disoriented by the experience.

Though MIT famously experienced an institutional crisis between 1968 and 1971, its elite status protected it from a major decline in engineering enrollment, tenured faculty, or federal funding. On the contrary, it was among the largest recipients of contracts from the Ford Foundation, the NSF RANN program, and the like.[74] Nonetheless, as the preeminent site for imagining the future of engineering, debates about sponsored research, the value of instrumental knowledge, and the nature of technology spurred pedagogical reform that embodied conflicting combinations of the innovations at southern California's engineering colleges. MIT emerged from its institutional revolt with ambitious plans for a new kind of hybrid expert.

Understanding MIT's curricular innovations requires returning again to the aftermath of World War II. No institution experienced a transformation that matched MIT's, whose Radiation Laboratory pioneered large-scale interdisciplinary research in the nation's service. In 1947, under the direction of chemical engineer Warren K. Lewis, the Institute undertook a two-year survey to determine how to create leaders for the world it had built. The Report stressed the magnitude of MIT's metamorphosis—captured by the fact that sponsored research now surpassed the academic budget—and concluded that there was "no road back" from modernity's epochal changes. It recommended an "integral plan" in which engineering, science, and the humanities were mutually reinforcing.[75] Humanistic-social study would account for 20 percent of coursework in a new School of Humanities—divided into departments of Economics and Social Science, English and History, and Foreign Language. Education in the arts, humanities, and social studies would develop leadership skills, reduce graduates' feelings of cultural inferiority, and attract students from elite prep schools.

In its first decade, the School of Humanities evolved according to a different logic than its intended mission of general education.[76] Economics offered its first doctoral training in 1941, swiftly achieving international standing. Graduate programs in Political Science, Linguistics, and Psychology followed, and the School changed its name to Humanities and Social Sciences. Faculty emulated colleagues elsewhere at MIT, pursuing sponsored research and goal-oriented projects. This research orientation left the historians and literature professors in a renamed Humanities department responsible for general education. In 1955, Humanities created an integrated major known as Course XXI that required students to balance their training between a field of engineering and of humanistic study to produce scientifically literate administrators and executives.[77]

As MIT's social scientists and humanists diverged, the School of Engineering overhauled its curriculum to keep pace with postwar growth. In an environment of expanding research dollars, engineers were alarmed as enrollment in their majors declined by 8 percent between 1952 and 1962 while science majors increased by 113 percent.[78] Under the direction of dean Gordon S. Brown the school received an unprecedented $9 million award from the Ford Foundation to fund seven endowed chairs and revamp the undergraduate curriculum with a foundation in the engineering sciences and student research. The School replaced outmoded laboratories and techniques and produced a generation of engineering science textbooks that influenced pedagogy worldwide.[79]

The engineering science curriculum was premised on interdisciplinary exchange with the natural and physical sciences, but there was scant mention of the humanities or social sciences. By the mid-1960s, however, the liberal qualities of the ASEE's "genuine engineer" received renewed attention at MIT. In 1965, the Department of Humanities won a $250,000 Carnegie Corporation grant to re-evaluate its mission in a series of summer meetings at Endicott House that concluded in the 1968 meeting where participants predicted societal breakdown in the absence of curricular reform.[80]

Indeed, MIT's troubles grew in the fall of 1968 with student protests, escalated with the March 4, 1969 work stoppage, and climaxed in the winter of 1970 with the student occupation of the President's office. Yet while MIT struggled with the politics of the military-industrial complex, curricular innovations flourished as a means of addressing the unrest. With the aid of new grant sources, faculty established centers for sociotechnical research such as the Urban Systems Laboratory, the Sea Grant Program, and the Cambridge Project. In every School at MIT, new courses sought to understand technology in its social context. Enrollment in Humanities and Social Sciences surged to two hundred majors in 1969–1970, twice the enrollment from four years earlier.[81] Leftist faculty, including linguist Noam Chomsky, historian Richard Wertz, psychologist Stephan Chorover, historian of science Nathan Sivin, and mechanical engineer Thomas B. Sheridan successfully created the Program on Social Inquiry, designed to instill "constructive social change" through a balanced distribution of training in critical theory and an engineering or science field.[82]

In 1970, MIT president Howard Johnson established a commission charged with examining MIT's culture in its entirety. Its report, *Creative Renewal in a Time of Crisis*, described a student's experience as specialized and fragmented. Bolstering its findings with insights from Roszak, Arendt, and Goodman, it reiterated the goal of liberally educated professionals and suggested sweeping changes so that students and faculty alike would recognize the normative character of their work. All faculty were to take responsibility for a students' first two years. An organizational structure known as the First Division should be created to administer the integration.[83] The Commission's minority report, however, doubted such "humanizing" could be achieved without a fundamental integration of value judgment in technical coursework.[84] Indeed, while innovative pilot projects such as the Concourse Program aspired to integration, a First Division never took hold.

A series of administrative transitions from the President's office to the department level nevertheless held on to the idea of integrating social and technical knowledge. Jerome Wiesner, who replaced Johnson as president in 1971, emphasized a new breed of socially and politically savvy technologist. When Wiesner returned to MIT in 1966 as dean of science, then provost, and finally president, he sought to incorporate policy-making into the Institute's mission. He envisioned MIT as an incubator of technical leaders who would serve civil society with results that matched their contribution to national defense.[85] His vision was heavily informed by experience. Early in his career he had served on the Lewis Committee. He went on to become a science advisor in the Eisenhower, Kennedy, and Johnson administrations. He also was a longstanding patron of the arts who had helped make the art/science/technology Center for Advanced Visual Studies a reality. As early as 1967, Wiesner discussed the possibility of building an experimental college at MIT that integrated science, engineering, arts, humanities, and social sciences.[86]

Alfred A. H. Keil, the new dean of engineering under Wiesner, expressed a similar vision. Anticipating socio-technical consequences would require engineers to become "multidisciplinary generalists" who understood the "interaction between advancing science and technology and the development of society."[87] This suggested a reorientation of the School of Engineering's research agenda to "operating systems of society" and the development of liberal education. Still, Keil admitted, in an institution made up of the world's leading specialists, it would be a challenge to train socio-technical generalists and even harder for "broad systems engineers" to get tenure.[88]

In addition to MIT's administration, a group of over fifty professors and students known as the Technology and Culture Seminar offered yet another model of technology & society education. In 1971, Reverend John Crocker Jr., MIT's Episcopalian minister led the Seminar to restore MIT's "normative guidance."[89] Participants included Wiesner, Keil, Hollomon, and the department heads of Aeronautical, Mechanical, and Electrical Engineering; Mathematics; Physics; and Political Science; but also activist faculty including Sheridan, Chorover, Oglesby, Joseph Weizenbaum, and Salvador Luria. As the Seminar grew, it sponsored public lectures by speakers ranging from Mumford and Illich to Jay W. Forrester and Daniel Bell.

The Technology and Culture Seminar's major work took place in weekly luncheons. These discussions drew out an awareness of fundamental differences in ideology that infused pedagogical reform at MIT. Disagreement hinged on competing visions of societal change and the role of values in design. In one meeting, for example, the radicalized mechanical engineer Thomas B. Sheridan brought Ellul's *Technological Society* to initiate a discussion about engineers' moral responsibility. Louis D. Smullin, department head of Electrical Engineering, argued in response that individual engineers could not be responsible for the minute effects of individual decisions, and that problems could only be addressed from a systems perspective.[90]

Wiesner's vision, combined with pockets of support across MIT, coalesced in a plan to establish a new field of Technology Studies. Ambitions ranged from a small campus center to a "unit of type unprecedented by the Institute."[91] A steering committee chaired by historians Harold J. Hanham and Elting E. Morison formed in 1973 that consisted of five historians, four engineers, a physicist, and a biologist.[92] Most were regular participants in the Technology and Culture Seminar. Emphasizing the normative nature of all engineering tasks, they defined their work in opposition to the "value-free model-building studies of technical change usual in the social sciences."[93] "The issue is responsibility," Sivin wrote, "which no systems-analysis technique will ever redeem from being a matter of individual moral choice."[94]

The nascent program identified the undergraduate education of scientists and engineers as its *raison d'être*. Without shared instruction by engineering faculty, humanistic courses would "recede into the background noise of the undergraduate's four years of institutionalized distraction."[95] Aerospace engineer Louis Bucciarelli in particular strove to bridge the divide. He argued that though the critique proffered by theorists of technological politics took "a very broad perspective on its problems and is overdrawn," it provided a window into design alternatives.[96]

Technology Studies became an official program in 1975 with a core faculty of seven. Bucciarelli and nuclear engineer Irving Kaplan were the program's advocates in the School of Engineering, and the political scientist Langdon Winner and oral historian Charles Weiner were hired to help build the program on the side of the humanities and social sciences. Technology Studies offered courses on the "Social Responsibility of the Scientist and Engineer," "Theories of Technological Society and Politics," and "Aesthetics in Science and Technology."[97] It had no formal control over professional seminars, however, which were initiated at the discretion of participating departments. Thus, despite its innovative courses, the Program never achieved broad integration with the engineering curriculum.

By the time Technology Studies took shape, moreover, MIT's institutional crisis had cooled. Student disruptions abated along with fears of an economic collapse. At MIT the number of students served by Humanities and Social Science electives declined and Course XXI majors plummeted to pre-crisis levels by 1975. After 1972 all MIT undergraduates had to focus their humanistic-social electives in a concentration. In its best year Technology Studies had 27 concentrators out of 1812 students.[98]

Plans for creating socio-technologists, however, had administrative momentum. In the second half of the 1970s, MIT raised $4 million from the Sloan, Mellon, Flora Hewlett, Max Fleischmann, and Exxon Foundations with the intention of building Wiesner's integrated college. The project resurrected longstanding ideological battles at MIT and beyond, with the journal *Nature* characterizing the effort as "a second order 'reaction to the public reaction'" to technology's critics.[99]

In the early 1980s STS at MIT became professionalized. It co-sponsored the journal *Science, Technology, and Human Values* with Harvard's Kennedy School. It found greatest student interest not among engineering undergraduates and the faculty that wished to instruct them, but among graduate students in the humanities and social sciences. Though the size and quality of its faculty in history, sociology, and philosophy grew rapidly, STS lamented the increasing absence of scientists and engineers affiliated with the Program.

MIT's effort to create socio-technologists, thus presents an ambivalent legacy. Throughout the 1970s less than 5 percent of MIT undergraduates had taken a class in Social Inquiry, Technology Studies, or STS.[100] One of the Institute's great strengths was the relative independence of its units from each other, but that independence worked against the integration that plans called for. In the aggregate, MIT's engineers of the 1970s were not a "new breed" apart from those of the 1960s. The engineering sciences, funded by sponsored research for military applications, remained the norm. The "relation of knowledge to values" advocated in *Creative Renewal in a Time of Crisis* had its proponents, but absent a First Division it did not become MIT's guiding philosophy.

Conclusion

In the two decades following World War II, engineering education in the United States underwent a contentious but enormously successful reorientation to engineering science. Funding and prestige followed a vision of the scientific engineer that was "maximally appropriate" for its time and space.[101] When the engineering science model threatened to collapse from internal and external pressures, reformers challenged local and national institutions with new ideas of who engineers should be. The forms of knowledge they advocated were not solid-state physics or multivariable calculus, but rather studies of the urban environment and the social theory of technology.

Variants of the pedagogical innovations and tribulations at MIT and Southern California's technical schools could be found in almost any engineering college in the United States. Most replaced general education with liberal learning. Most created extracurricular programs in the arts. Most experimented with a technology & society course designed in tandem with humanists and social scientists. A smaller number—including Cornell, Carnegie Mellon, Lehigh, Penn State, Stanford, and Virginia Tech—pursued new technology studies programs.[102] All dealt with the question of whether technology was to blame for societal crisis.

Despite similarities, technology & society pedagogy could encompass a range of assumptions and ambitions about the engineer's societal role. UCLA's reformers wanted engineers to be expert managers of the public good. Caltech's social scientists argued that they, not scientists or engineers, should manage social progress. HMC was founded on the belief that engineers would shape social change; it came to instruct, however, that engineers should be introspective about the limitations of changing the human condition through technology. MIT pursued elements of all of these plans as it tried to create a new college for socio-technologists.

Conceptions of agency and control were guiding factors in curricular reform. In some cases engineering educators appropriated social theoretical texts that emphasized the primacy of values and norms over autonomous change. They argued that technological fixes could have diminishing returns and that problem definition needed to integrate myriad social dimensions. The great majority of educators, however, implicitly or explicitly conveyed an ideology of technological change. This was the discursive frame that UCLA used to conclude that "the rising tide of social and technological change" had "engulfed" the professions.[103] It was the basis of "Science for Mankind" and Caltech's social sciences. And it was the position from which HMC started: "The fundamental fact of modern life is the acceleration of change," Platt asserted, "This fact is so obvious as not to require documentation."[104]

Ultimately, however, a systemic transformation to the education of comprehensive socio-technologists of any sort eluded the nation's engineering schools. In 1973, MIT's Center for Policy Alternatives offered a series of prescriptions for integrating a "quantitative problem-solving orientation" with "knowledge of the collective values, institutions and cultural patterns of society." Design training should begin in a student's freshman year and carry throughout the curriculum; all students should have project-based clinical experience; faculty across the university needed to take collaborative responsibility for a student's "total education"; and an engineering curricula ought also to be a liberal education for professionals in other fields.[105] But the report, chaired by Hollomon, expressed frustration over the lack of progress. "A decade of protest and turmoil has left engineering campuses and students only slightly changed," Hollomon wrote. "As before, the engineering student is comparatively pragmatic, self-directed, not people-oriented, and desirous of unambiguous situations and structured work."[106]

Hollomon's lament signaled the beginning of a precipitous decline in efforts to make socio-technical problems central to engineering pedagogy. Universities and technical institutes were left with STS programs, minor increases in humanities curricula, and brochures with an aura of human values. However, these took their place with past reforms: institutionalized, but on the outskirts of engineering culture.[107] The economic recession of the early 1970s made resources for innovative programs scarce, while tenure structures attenuated commitment to pedagogical change. Efforts to raise an institution's technical reputation competed for funding with social-action programs to the detriment of the latter.[108] Moreover, despite rare individuals like Rhodes at Caltech, even at the height of the student movement most incoming students already internalized the division. If faculty could not sustain reflective integration of "social" and "technical" knowledge, how could they expect future generations of engineers to do so?

8

Epilogue

Prometheus the creator, once restrained by defense projects sharply focused upon technical and economic problems, is now free to embrace the messy environmental, political, and social complexity of the postindustrial world.
—Thomas P. Hughes, *Rescuing Prometheus*[1]

When aspiring engineers asked what their careers would entail in 1975, they were confronted with a revised future from a decade prior. In the second edition of the *Engineer and His Profession*, John Dustin Kemper surveyed the rise and fall of the space program, the Vietnam War, and the counterculture to explore what had changed. First and foremost, the megamachine had not taken over and its prognosticators were on the wane. Closer to home, reformers such as the Committee for Social Responsibility in Engineering did not rally colleagues to political action in great numbers. The decline of "extreme advocacy" in engineering was for the better, Kemper argued, because "to assume a hard-line moral position which moves very far ahead of public opinion" was a path to technocracy.[2] Nonetheless, the American public no longer shared a universal devotion to technology. Engineers faced an environment in which they were berated for causing problems while simultaneously called upon to provide solutions. In an expansive addition to his textbook, Kemper surveyed the state of the art in ethics, the environment, and energy. The problems were enormous and politics were inevitable, he concluded, but engineers would overcome them.[3]

Reformers in the 1970s echoed Kemper's list of challenges but continued to stress the potential of alternative paths. Investigations into social responsibility flourished on college campuses.[4] STS programs drew pockets of introspective engineers to question the boundaries between the "social" and the "technical." The creation of the EPA gave rise to environmental engineering as a new regulatory field at the nexus of government, academe, and industry. The appropriate technology movement evolved into enterprise organizations. Hobbyist clubs, artist collectives, and corporate research laboratories fueled the development of personal computers by stressing access and usability.

In an environment of stagflation and energy crisis, awareness of the socio-technical dimensions of engineering had become nearly universal. However, alternative visions of technology maintained a partisan and marginal cast. Reformers still found critical social texts vital, but to argue that technical problems were inherently political remained controversial. Engineers in the EPA, for example, came to be defined as technical support rather than proactive designers of environmental protection. The appropriate technology movement remained dependent on volunteer efforts and government funding, and struggled with conflicting values of expertise and community partnership.

If environmental engineers and appropriate technologists did not produce an ecological revolution, they fared better than the profession's peace advocates. The social movements of the 1960s and 1970s introduced thousands of engineers and students to public service and social entrepreneurship. Military research was converted to civilian ends on a small scale, particularly through grants made available for applying systems techniques in urban environments. However, critiques of militarism among engineers diminished as the United States withdrew from Vietnam, the aerospace industry bottomed out, and energy and environment became problems in need of high-tech solutions. By the early 1980s engineering enrollments rose in tandem once again with increased defense spending, limiting the appeal of the reformers' message.[5]

Reformers discovered moreover that an ideology of technological politics raised more questions than it answered. Rather than normalizing *ought* into *is*, critical theorists emphasized the divide between the two and offered little guidance on how to bridge it. They charged that engineers were complicit in the destructive and dehumanizing values of society, that engineers did not control the direction of their labor, that attempts to convert their work to domestic problems could not escape the system's totality. New social texts and modes of organization offered the hope of fulfilling work and a clean conscience, but to follow through could mean rejecting economic stability, colleagues, and the myriad comforts of the status quo. It also created disagreements between radicals and reformers about expertise, status, and professional identity. Should technology be remade outside the constraints of formal expertise through democratic design? Inside the corporation where engineers had responsibility for foreseeing the impacts of their labor? By merging a historical ideal of engineers as independent public servants with rigorous social-theoretical training? The diversity of opinion on what constituted engineering service limited reformers' ability to enroll rank and filers much less to convince superiors.[6]

An ideology of technological change, however, became a naturalized theory for survival in a global society, dissociated from the debates that brought it to prominence. In the 1980s a "doctrine of competitiveness" dominated visions of technology and sparked national initiatives from professional societies, engineering educators, and advisory groups.[7] The nation, the argument went, was falling behind due to a lack of economic productivity; the solution was to invest in technology and roll back environmental regulations. The framers of the competitiveness doctrine were in many cases the same visionaries who had waged war with anti-technologists. In his 1981 book *America's Technology Slip*, for example, Simon Ramo reiterated his argument about a mismatch between technological change and social institutions, but now stressed that the failure of the nation's educational and government policy was having an adverse effect on the ability of the United States to keep pace with global competition.[8] This view was bolstered by managerial innovations such as total quality management and "re-engineering."[9]

Education remained a key site of contention for defining who engineers ought to be. Prominent engineers, corporate executives, government officials, and social scientists again decried the engineering profession's public image, technical skills, and leadership qualities. A series of reports published by the National Research Council (NRC) in 1985 offered policy prescriptions so that America's educational system could keep pace with converging historical forces. Its investigation identified four radical changes stressing engineering labor. The first was an unprecedented expansion of government. Federal intervention in the economy had fueled new specializations such as aerospace engineering in the 1950s, nuclear engineering in the 1960s, and environmental engineering in the 1970s but neglected private sector growth. The second disruption was an "information explosion" that surpassed even the steam engine in historical significance, broadening engineering into realms of software. The third was the general acceleration of technological change, the rate of which prompted the panel to doubt that "the engineering supply system" could "continue to adapt."[10] Finally, the worldwide expansion of American corporations had sowed the seeds of international competition. The NRC's underlying ambition was to study the "mechanisms and limits of change" so that "informed choices" would result in mastery rather than a "crisis-response posture." It stressed the necessity of "managing change" to cope with job dislocation and obsolescence. Disruptions caused by rapid change were of paramount importance; however, the rank and file should stay on task, assured that technology's social impacts were "a management problem and a political problem."[11]

The ubiquity of an ideology of technological change has since benefited from new rhetorical innovations. From the 1980s through the dot-com boom, "information technology" moved from cybernetic theorists and management consultants into the international popular lexicon.[12] Similarly, when "globalization" emerged as a keyword in the late 1980s, its evangelists embodied the tenets of an ideology of technological change. To read Thomas Friedman's ten "flatteners," is to encounter inevitable acceleration in a new register. Singling out engineers, Friedman implores that: "There is no sugar-coating this: in a flat world, every individual is going to have to run a little faster."[13]

Taken together, then, when we look across engineering cultures from the 1960s to the present, we see on the one hand the emergence of alternative visions of technology that struggle to gain traction but nonetheless continue to draw an eclectic minority of reformers. On the other, an ideology of technological change is an almost universally accepted description of reality. Engineers were not the sole creators of this vision of inevitable acceleration, but they were among its greatest advocates. Whatever their limitations as social theorists, Ramo, Mesthene, and the ideologues of change described a revolution that has come to pass.

Revisiting the conflicts of the late 1960s from the vantage of contemporary debates about the status and obligations of engineers, however, induces a feeling of déjà vu. In one of the few synthetic accounts of postwar engineering, Thomas P. Hughes expresses optimism about engineers' ability to solve America's environmental, infrastructural, and energy challenges. Hughes argues that the Atlas missile program, the Semi-Automatic Ground Environment, and other military-industrial projects have given way to a social-industrial complex informed by collaborative, nonhierarchical methods. He concludes that engineers have been emancipated in a "postmodern" environment of horizontal organizations defined by interdisciplinarity, distributed control, networked systems, and continuous change, in which the Internet is the ultimate example.[14] The growth of green engineering, human–computer interaction, and sustainable design gives further credence to this perspective. So does the investment in hybrid and electric vehicles and alternative energy sources by entrepreneurs, venture capitalists, and the world's largest automakers.

Yet, when mission statements such as the ASCE's *The Vision for Civil Engineering in 2025* echo aspirations from three-decades prior, when socio-technical pedagogy continues to vex educators, when "greenwashing" requires little explanation, when Predator drones redefine asymmetric warfare, the criticisms of Paschkis, Slaby, and the CSRE resonate.[15] Given that American engineering remains oriented to service in large corporations and that the most important emergent fields such as robotics and nanotechnology are supported overwhelmingly by defense contracts, one might accept Noble's less sanguine reading of engineering history. "Modern Americans," Noble writes, "confront a world in which everything changes, yet nothing moves."[16]

We need adopt neither Panglossian optimism nor stasis, however, to point out that the human-built world has improved since the height of the cold war and that cultural politics continue to challenge engineers. Within the past decade the engineering profession has experienced dramatic cultural and intellectual fervor in rekindled debates about responsibility, service, and creativity. In an environment of globalization, climate change, terrorism, and controversial wars, reformers have sought to remake engineering for a networked, global economy and have added "sustainability," "social justice," and "global competencies" to the engineer's vocabulary.

A small group of dissatisfied engineers have again embraced critical theory to challenge professional norms. Largely in response to the Iraq and Afghanistan wars, these groups have reiterated critiques of the late

1960s while calling for a rejuvenated engineering identity around the notion of responsibility and social justice. At MIT, for example, students introduced a graduation oath that states: "I pledge to explore and take into account the social and environmental consequences of any job I consider and will try to improve these aspects of any organization for which I work."[17]

The most explicit vision of alternative engineering service is found in a loose affiliation of professors, artists, and students known as the Engineering, Social Justice, and Peace (ESJP) network. Founded in 2004, ESJP is led by a coordinating committee that includes George D. Catalano a professor of biotechnology at SUNY Binghamton and former instructor at the United States Military Academy, Caroline Baillie of the University of Western Australia, and Donna Riley of Smith College.[18] Their vision of social justice is based on liberation theology, postcolonial theory, and critiques of neoliberalism. Issues of the ESJP "zine" *Reconstruct* print conversion stories, critiques of weapons systems, as well as poetry and artwork with the ambition of using "critical analyses to *deconstruct* engineering-as-usual, and to examine social *constructs* of engineering that perpetuate a culture of militarism and materialism."[19] ESJP participants are, in the main, academics who have published a series of "synthesis lectures" on topics such as *Engineering Ethics: Peace, Justice, and the Earth* and *Engineering and Society: Working toward Social Justice*, which are distributed in print and as electronic textbooks.[20] Aimed at an undergraduate and working audience, they offer alternative problem-defining frameworks that push readers to recognize the myriad social factors in any technical problem. The ESJP network, like its counterparts in the 1960s, constitutes a tiny fraction of all engineers, but their framing of social justice has been taken up by the National Academy of Engineering in a series of workshops bringing ESJP members into contact with national professional leaders.[21]

More successful than the critical theorists has been the rapid growth of volunteer development projects and engineering NGOs. The most heralded is Engineers Without Borders (EWB-USA), which formed in 2002 out of a student project directed by University of Colorado, Boulder professor Bernard Amadei. EWB-USA now has over 12,000 members in 250 local chapters who work in 45 countries on water and sanitation projects. EWB-USA has attracted students and sponsors by downplaying

12   (re)construct, v.3

On April 24, 2011, I participated in the Peace Pilgrimage for a Nuclear Free World. The Peace Pilgrimage, organized by Jun-san Yasuda from the Grafton, NY Peace Pagoda, began April 10, 2011 with a vigil at the Indian Point Nuclear Power Plant, with walkers making a 206 mile trek to Vermont Yankee, where they held another vigil. Although about 75 people participated in the last leg of the trek from Marlboro and Brattleboro to the northwest, the march I participated in came from the southeast, in my home state of Massachusetts.

My reasons for marching were made clear on my signage. The first principle in safety engineering is to ask yourself "Can you design out the risk?" And if the answer is yes, then you do so. What Fukushima and the 25th anniversary of the Chernobyl disaster have brought into stark relief is the fact that engineers cannot anticipate everything that can happen, and risks we had quantitatively estimated as incredibly remote are happening with higher frequency than predicted. To design for the unanticipatable, we must think holistically about the scope and scale of our activities and their risks — and design out the environmental and human disasters that have become business as usual.

As folks in the US have revisited our own history of nuclear accidents, I am reminded that the largest release of radioactivity was not at Three Mile Island as is popularly assumed, but on Navajo lands at Church Rock, New Mexico. A dam failed at the site of a uranium mill there in 1979, resulting in a release of radioactive waste and heavy metals into the Puerco River, where contamination continues. While these long-lasting environmental and human costs of nuclear power are not widely known, they ought to have central consideration for those of us working for social justice.

~Donna Riley (excerpt from esjp.org)

Figure 8.1
Engineers, Social Justice, and Peace
Source: Donna Riley, "Peace Pilgrimage for a Nuclear Free World," *Reconstruct: A zine about Engineering, Social Justice, and Peace* 3 (2011): 2. Courtesy of Usman Mushtaq and Donna Riley.

political ideology and at the same time drawing on progressive era identifications of the engineer, attracting grants and donations from the ASCE,
ASME, environmental design firms, as well as Boeing and Chevron.[22]
This big tent philosophy has been productive in drawing attention to
alternative visions of engineering service and in enrolling students as
its agents and ambassadors. Like earlier appropriate technology efforts,
however, it is torn between putting technical products at the center and
balancing the social and cultural dimensions of community development.
The reliance on student volunteers, which is the organization's strength,
also raises concerns about "voluntourism."[23] Nonetheless, EWB-USA and
other groups such as Engineers for a Sustainable World are altering
notions of engineering to a degree comparable to the reformers of the
1960s.[24]

Linking intellectual and political themes from outside of traditional
practice has been crucial for re-contextualizing new forms of service and
problem definition. *Sustainability*, for example, is a theme that cuts across
government, corporate, and academic audiences, which, like responsibility, serves as a boundary concept to rethink problem-solving methods.[25]
One of the successes in this regard is the Colorado School of Mines'
Humanitarian Engineering minor, founded in 2003. Directed by mechanical engineer David Muñoz, the Humanitarian Engineering program
received a $1.1 million grant from the Hewlett Foundation to integrate
humanities and social sciences through courses in engineering cultures,
development policy, energy economics, anthropology as well as subjects
like control systems and wastewater engineering. Students are required
to contribute to development projects, often through EWB-USA.

Promising as the proliferation of alternatives has been, their integration into the engineering mainstream has encountered the same hurdles
as their predecessors in the late 1960s. Humanitarian Engineering, for
instance, remains an undergraduate minor and is cast in the press as "the
softer side of engineering," of lesser significance to the techniques of
scientific theory.[26] Moreover, though the minor promotes the engineer as
a change agent able to "design under constraints to directly improve the
well-being of underserved populations," it contends also that this will
benefit traditional engineering labor patterns through a general enhancement of "sociocultural awareness."[27]

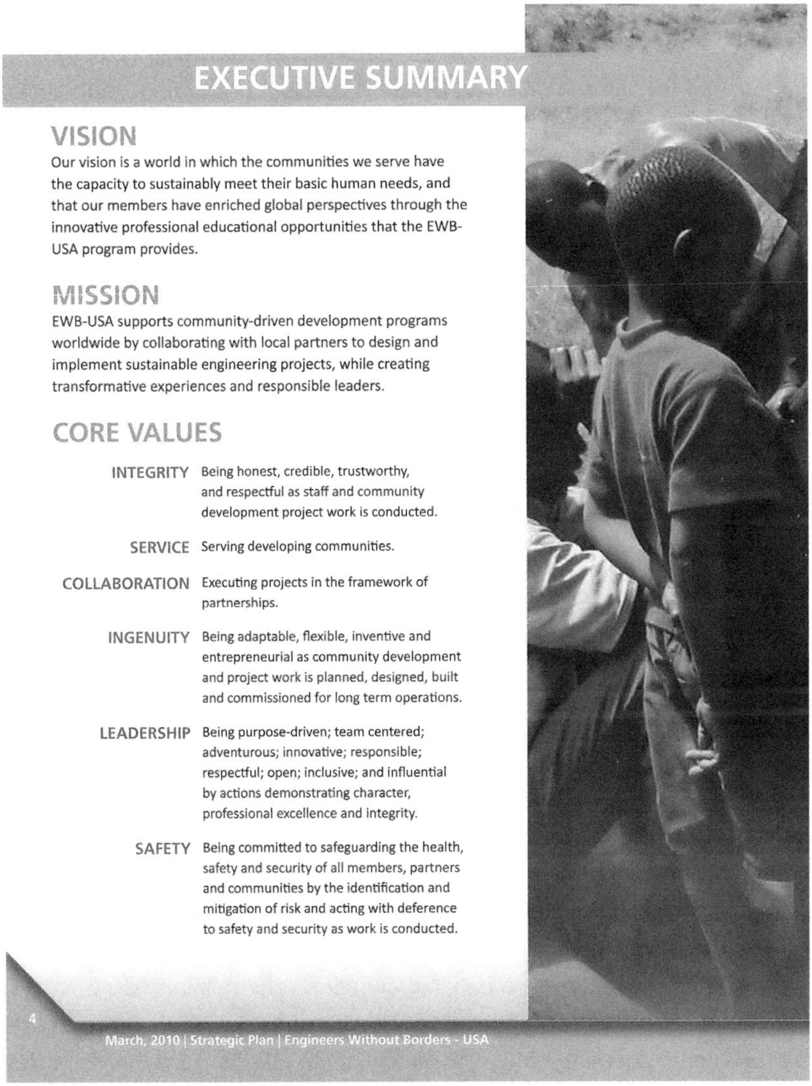

EXECUTIVE SUMMARY

VISION

Our vision is a world in which the communities we serve have the capacity to sustainably meet their basic human needs, and that our members have enriched global perspectives through the innovative professional educational opportunities that the EWB-USA program provides.

MISSION

EWB-USA supports community-driven development programs worldwide by collaborating with local partners to design and implement sustainable engineering projects, while creating transformative experiences and responsible leaders.

CORE VALUES

INTEGRITY — Being honest, credible, trustworthy, and respectful as staff and community development project work is conducted.

SERVICE — Serving developing communities.

COLLABORATION — Executing projects in the framework of partnerships.

INGENUITY — Being adaptable, flexible, inventive and entrepreneurial as community development and project work is planned, designed, built and commissioned for long term operations.

LEADERSHIP — Being purpose-driven; team centered; adventurous; innovative; responsible; respectful; open; inclusive; and influential by actions demonstrating character, professional excellence and integrity.

SAFETY — Being committed to safeguarding the health, safety and security of all members, partners and communities by the identification and mitigation of risk and acting with deference to safety and security as work is conducted.

March, 2010 | Strategic Plan | Engineers Without Borders - USA

Figure 8.2
Engineers Without Borders–USA
Source: EWB-USA, *Strategic Plan* (Boulder, CO: EWB-USA, 2010), 4. Courtesy of EWB-USA.

Alongside social justice and sustainability movements, there has been a resurgence of a vanguard that believes technological change is forcing an epochal transition. The Acceleration Studies Foundation, founded in 1999 by the futurist John Smart, for example, claims over three thousand members including prominent engineers, science fiction authors, entrepreneurs, corporate executives, and NASA scientists. Its website Acceleration Watch provides links to its Future Salon Network and information on leading educational programs for producing "change managers," mostly located in technology-oriented business schools, which it contrasts with "the 'cultural relativist' . . . dead-end narcissism and nihilism" of STS programs.[28] The lead proponent for mastering epochal change has been artificial intelligence pioneer Ray Kurzweil. In a series of books and interviews he foretells of the coming "singularity" where rapid technological advances create a transhumanist society.[29] In 2007, along with Peter Diamandis, he founded Singularity University on NASA's Ames campus. Among its instructors is Google Internet evangelist Vint Cerf and Ethernet inventor Bob Metcalfe. In 2010, Singularity University described its mission as the identification and use of "exponentially-accelerating technologies to create better conditions for everyone on earth; to heal and nurture the planet itself; and to guide humanity as it reaches beyond the limits of the Earth."[30] As it was for the Innovation Group, this is a proactive effort to shape the future, leveraged as a philosophy of personal adaptation. A recent executive education brochure declares: "Be prepared to learn how the growth of exponential and disruptive technologies will impact your industry, your company, your career, and your life."[31]

Engineers not only presided over change in the postindustrial world, they contextualized it in powerful frames that realigned the meanings of *technology* and *engineering*. The vision of change they embraced points us to the ways in which the language of technology can be especially problematic for engineers. Neither an ideology of technological change nor technological politics was a natural choice for the profession. The concept of technological change, however, at least shared many of the premises of engineers' existing technocratic ideology. Its proponents asserted that technology was an inevitable force in society, that its control necessitated their expertise, and that with an assertion of social responsibility engineers might attain a role as policy makers. Additionally such

"Be prepared to learn how the growth of exponential and disruptive technologies will impact your industry, your company, your career, and your life."

Bob Metcalfe
Founder of 3Com
Co-inventor of the Ethernet
General Partner, Polaris Venture Partners

Figure 8.3
Singularity University
Source: Singularity University, *Exponential Technologies Executive Program* (Moffett Field, CA: Singularity University, 2011), 3. Courtesy of Singularity University.

a theory explained away serious criticisms of technology as manifestations of culture lag. Most critically, it posited that technology's ill effects were not the result of engineering failures; rather, those effects derived from the nature of technology. But this vision undermined engineers' claim to unique agency in the production of technology. It suggested that engineers were another element of society that needed to keep pace. An ideology of technological change provided a solution to the crisis of technology while undercutting the identity of the engineer.

Faced with a plethora of challenges, mainstream engineering institutions from the professional societies to the NAE and the technical universities continue to rely on this vision at the same time they seek to overcome its logic of obsolescence. The NAE's most recent reform effort, *The Engineer of 2020*, portrays engineers as faced with becoming either technology's victims or its masters. "We must ask if it serves the nation well to permit the engineering profession and engineering education to lag technology and society," the executive summary reads, "especially as technological change occurs at a faster and faster pace. Rather, should the engineering profession anticipate needed advances and prepare for a future where it will provide more benefit to humankind?"[32]

It is my hope that this account of competing visions of technology proves valuable not only to scholars of engineering studies but also to contemporary reformers, whether in ESJP, the NAE, or simply in the classroom or the design office. At the height of technology & society debates, engineers were among the most reflective thinkers about the possibilities and limits of their interventions. Participants recognized that engineering is an inherently normative process and that ideological frames matter in how engineering problems are defined, even which problems are deemed worthy of attention. For engineers and those who teach their future ranks, revisiting this process of contextualization is important, not because it proscribes a new way of life, but because it insists our assumptions remain perpetually contested. That, after all, is the basis for change.

Notes

Chapter 1

1. Quoted in Carlo Ginzburg, *The Cheese and the Worms: The Cosmos of a Sixteenth-Century Miller*, trans. John Tedeschi and Anne Tedeschi (Baltimore: Johns Hopkins University Press, 1980), 13.

2. CSRE, "Packard???" *Spark* 1, no. 1 (spring 1971): 46–47.

3. "Engineering Crisis" flyer, 1971, box 4, Miscellaneous, Victor Paschkis Papers, Swarthmore College Peace Collection [hereafter cited as Paschkis Papers].

4. CSRE, "Packard???" 46.

5. Stephen H. Unger, "New Engineering Conference: CSRE holds Counter-Conference during IEEE Convention," *Spark* 1, no. 2 (fall 1971): 2–5.

6. Nat Snyderman, "Protesters Grill IEEE Speakers," *Electronic News* 16 (March 29, 1971): 19.

7. "HP History." Accessed February 17, 2012. www.8.hp.com/us/en/hp-information/about-hp/history/hp-timeline/hp-timeline.html.

8. UPI, "2 Firebombs Hurled at Home of Industrialist," *Los Angeles Times* (January 12, 1971), A2.

9. Snyderman, "Protesters Grill IEEE Speakers," 19; Institute of Electrical and Electronics Engineers, *1971 IEEE International Convention Digest* (New York: IEEE, 1971), 230.

10. Clarke A. Chambers, "The Belief in Progress in Twentieth-Century America," *Journal of the History of Ideas* 19, no. 2 (1958): 197–224; Paul S. Boyer, *By the Bomb's Early Light: American Thought and Culture at the Dawn of the Atomic Age* (New York: Pantheon Books, 1987); Lizabeth Cohen, *A Consumers' Republic: The Politics of Mass Consumption in Postwar America* (New York: Vintage, 2003); Megan Prelinger, *Another Science Fiction: Advertising the Space Race, 1957–1962* (New York: Blast Books, 2010).

11. Jacques Ellul, *The Technological Society*, trans. John Wilkinson (New York: Knopf, 1964), 14.

12. Herbert Marcuse, *One Dimensional Man: Studies in the Ideology of Advanced Industrial Society* (Boston: Beacon Press, 1964), 1–3.

13. Lewis Mumford, *The Myth of the Machine: Technics and Human Development* (New York: Harcourt, Brace and World, 1967), 3.

14. Theodore Roszak, *The Making of a Counter Culture: Reflections on the Technocratic Society and Its Youthful Opposition* (Garden City, NY: Doubleday, 1969), xii.

15. Edwin T. Layton Jr. *The Revolt of the Engineers: Social Responsibility and the American Engineering Profession* (Baltimore: Johns Hopkins University Press, 1986), 57.

16. Ronald R. Kline, "From Progressivism to Engineering Studies: Edwin T. Layton's *The Revolt of the Engineers*," *Technology and Culture* 49, no. 4 (October 2008): 1018–24; David F. Noble, *America by Design: Science, Technology, and the Rise of Corporate Capitalism* (New York: Knopf, 1977); Jeffrey Herf, *Reactionary Modernism: Technology, Culture, and Politics in Weimar and the Third Reich* (New York: Cambridge University Press, 1984); Kendall E. Bailes, *Technology and Society under Lenin and Stalin: Origins of the Soviet Technical Intelligentsia, 1917–1941* (Princeton: Princeton University Press, 1978); Dolores L. Augustine, *Red Prometheus: Engineering and Dictatorship in East Germany, 1945–1990* (Cambridge: MIT Press, 2007).

17. Donald E. Marlowe, "Review: *The Revolt of the Engineers*," *Mechanical Engineering* 93, no. 10 (October 1971): 66.

18. Samuel C. Florman, *The Existential Pleasures of Engineering* (New York: St. Martin's Press, 1976), ix. Florman described Layton's *Revolt of the Engineers* as one of his favorite books. Ibid., 155.

19. Ibid., 18.

20. These sentiments were particularly strong in the chemical engineering press. See, for example, L. H. Kaplan, "Letter to the Editor," *Chemical and Engineering News* 46, no. 18 (April 22, 1968): 6; Thomas Baron, "Changing Values and Our Industry," *Chemical Engineering Progress* 66, no. 12 (December 1970): 16–20; Fred L. Hartley, "Return to the Jungle?" *Chemical Engineering Progress* 67, no. 5 (May 1971): 34–36; Lauren B. Hitchcock, "Who Needs Engineers?" *Chemical Engineering Progress* 67, no. 7 (July 1971): 21–24.

21. Samuel C. Florman, "Anti-technology: The New Myth," *Civil Engineering* 42, no. 1 (January, 1972): 68–71.

22. B. L. McCorkle, "To the Editor," *Civil Engineering* 42, no. 4 (April 1972): 49; Sidney Zecher, "To the Editor," *Civil Engineering* 42, no. 4 (April 1972): 49.

23. Don Johnstone, "To the Editor," *Civil Engineering* 42, no. 4 (April 1972): 51, 59. See also Kim de Rubertis, "In Defense of Dubos and Ellul," *Civil Engineering* 42, no. 12 (December 1972): 50–53.

24. Robert E. Wakeman "Letter to the Editor," *Mechanical Engineering* 92, no. 11 (November 1970): 70.

25. Anonymous, *The Mute Engineers* (Princeton, NJ: Literary Publishers, 1974); Charles Susskind, *Understanding Technology* (Baltimore: Johns Hopkins University Press, 1973).

26. Kelly Moore, *Disrupting Science: Social Movements, American Scientists, and the Politics of the Military, 1945–1975* (Princeton: Princeton University Press, 2008).

27. Jennifer S. Light, *From Warfare to Welfare: Defense Intellectuals and Urban Problems in Cold War America* (Baltimore: Johns Hopkins University Press, 2003).

28. Fred Turner, *From Counterculture to Cyberculture: Stewart Brand, the Whole Earth Network, and the Rise of Digital Utopianism* (Chicago: University of Chicago Press, 2006).

29. Ken Alder, *Engineering the Revolution: Arms and Enlightenment in France, 1763–1815* (Princeton: Princeton University Press, 1997), 15.

30. Leo Marx, "Technology: the Emergence of a Hazardous Concept," *Social Research* 64, no. 3 (fall 1997): 965–88; Ruth Oldenziel, *Making Technology Masculine: Men, Women and Modern Machines in America, 1870–1945* (Amsterdam: Amsterdam University Press, 1999); Ronald R. Kline, "Constructing 'Technology' as 'Applied Science': Public Rhetoric of Scientists and Engineers in the United States, 1880–1945," *ISIS* 86, no. 2 (June 1995): 194–221; Eric Schatzberg, "*Technik* Comes to America: Changing Meanings of *Technology* before 1930," *Technology and Culture* 47, no. 3 (July 2006): 486–512.

31. Peter Meiksins, "*The Revolt of the Engineers* Reconsidered," *Technology and Culture* 29, no. 2 (April 1988): 219–46.

32. Ralph Landau, "The Chemical Engineer—Today and Tomorrow," *Chemical Engineering Progress* 68, no. 6 (June 1972): 9–19.

33. Ruth Oldenziel, "Signifying Semantics for a History of Technology," *Technology and Culture* 47, no. 3 (July 2006): 477–85.

34. Langdon Winner, *Autonomous Technology: Technics-Out-of-Control as a Theme in Political Thought* (Cambridge: MIT Press, 1977), 44–106; Rosalind Williams, *Retooling: A Historian Confronts Technological Change* (Cambridge: MIT Press, 2002), 14–19.

35. Simon Ramo, *Century of Mismatch: How Logical Man Can Reshape His Illogical Technological Society* (New York: David McKay, 1970), vi.

36. Joel Achenbach, "The Future Is Now," *Washington Post* (April 13, 2008), B1.

37. Historians of technology have touched on the issue, but as far as I know, no one has yet identified "technological change" as a term in need of a history. None of the authors in the anthology *Does Technology Drive History?* identify the concept of "technological change" as bearing specifically on the idea of determinism. Merritt Roe Smith and Leo Marx, eds., *Does Technology Drive History?: The Dilemma of Technological Determinism* (Cambridge: MIT Press, 1994). John M. Staudenmaier, in his monograph on the history of the SHOT, does not

identify "technological change" as an analytic construct with a particular history, despite his transformative encounter studying "the language of technological 'progress'" as a means of understanding its mythological properties. John M. Staudenmaier, *Technology's Storytellers: Reweaving the Human Fabric* (Cambridge: SHOT/MIT Press, 1985), xv. In their introduction to *Technologies of Power* Michael Thad Allen and Gabrielle Hecht point to the emphasis on technological change in the post-WWII era as a phenomenon closely tied to the cold war. Michael Thad Allen and Gabrielle Hecht, "Introduction: Authority, Political Machines, and Technology's History," in *Technologies of Power: Essays in Honor of Thomas Parke Hughes and Agatha Chipley Hughes*, Michael Thad Allen and Gabrielle Hecht, eds. (Cambridge: MIT Press, 2001), 1–23. Amy Sue Bix traces the trope of "technological unemployment," which is bound to the historical narrative of "technological change," but does not explore the meanings of "change" itself within employment debates. Amy Sue Bix, *Inventing Ourselves out of Jobs? America's Debate over Technological Unemployment, 1929–1981*, Studies in Industry and Society (Baltimore: Johns Hopkins University Press, 2000).

38. Marx, "Technology as a Hazardous Concept," 981–84.

39. Williams, *Retooling*, 17.

40. My interpretation of an ideology of technological change as a "dominant" vision is informed by a series of conversations with Gary Downey about the relevance of Antonio Gramsci's writings to engineering; Gary Lee Downey, "What Is Engineering Studies For? Dominant Practices and Scalable Scholarship," *Engineering Studies* 1, no. 1 (2009): 55–76.

41. Darrin M. McMahon, *Enemies of the Enlightenment: The French Counterenlightenment and the Making of Modernity* (Oxford: Oxford University Press, 2001).

42. Rick Perlstein, *Nixonland: The Rise of a President and the Fracturing of America* (New York: Simon and Schuster, 2008). For the impact of the culture wars in the 1970s and 1980s see Daniel T. Rodgers, *Age of Fracture* (Cambridge: Belknap Press of Harvard University Press, 2011).

43. James Davidson Hunter, *Culture Wars: The Struggle to Define America* (New York: Basic Books, 1991), 42–43. Hunter emphasizes that most religious culture warriors are not "intellectuals" in a formal sense and often lack "coherent, clearly articulated, sharply differentiated world views," but that they nonetheless have distinct moral visions. This analysis holds for many engineers in the rank and file in the 1960s. Engineering's reformers and elites, however, were seeking the "differentiated world views" of the intellectual.

44. Robert Zussman, *Mechanics of the Middle Class: Work and Politics among American Engineers* (Berkeley: University of California Press, 1985); Peter Whalley, *The Social Production of Technical Work: The Case of British Engineers* (Albany: State University of New York Press, 1986), 60; Edwin T. Layton Jr., "Veblen and the Engineers," *American Quarterly* 14, no. 1 (spring 1962), 70; John Rae, "Engineers Are People," *Technology and Culture* 16, no. 3 (July 1975), 404.

Chapter 2

1. Theodore Wachs Jr., *Careers in Engineering* (New York: Walack, 1964), 93–95.

2. Oldenziel, *Making Technology Masculine*, 51.

3. Samuel Smiles, *Lives of the Engineers, with an Account of Their Principal Works: Comprising Also a History of Inland Communication in Britain*, vol. 1 (London: Murray, 1861).

4. C. C. Furnas and Joe McCarthy, *The Engineer*, Life Science Library (New York: Time, Inc., 1966), 16.

5. Gene Marine, *America the Raped: The Engineering Mentality and the Devastation of a Continent* (New York: Simon and Schuster, 1969), 46.

6. Peter Meiksins, "Engineers in the United States: A House Divided," in *Engineering Labour: Technical Workers in Comparative Perspective*, Peter Meiksins and Chris Smith, ed. (London: Verso, 1996), 92; Monte A. Calvert, *The Mechanical Engineer in America, 1830–1910* (Baltimore: Johns Hopkins University Press, 1967), 197–224.

7. Meiksins, "Engineers in the United States," 61–97.

8. Gary Lee Downey and Juan C. Lucena, "Knowledge and Professional Identity in Engineering," *History and Technology* 20, no. 4 (December 2004): 393–420.

9. William H. Wisely, *The American Civil Engineer, 1852–1974: The History, Traditions, and Development of the American Society of Civil Engineers, Founded 1852* (New York: ASCE, 1974), 5.

10. Daniel Hovey Calhoun, *The American Civil Engineer: Origins and Conflict* (Cambridge, MA: Technology Press, 1960); Terry S. Reynolds, "The Engineer in 19th-Century America," in *The Engineer in America*, Terry S. Reynolds, ed. (Chicago: University of Chicago Press, 1991), 7–26.

11. Layton, *Revolt of the Engineers*, 3. Already in the 1840s the hierarchical structure of these new technological systems began to shape the patterns of engineering training and labor. Calhoun, *American Civil Engineer*, 182–99.

12. Noble, *America by Design*, 35–39.

13. Layton, *Revolt of the Engineers*, 25–43.

14. Noble, *America by Design*, 40–49; Oldenziel, *Making Technology Masculine*, 70–90; Layton, *Revolt of the Engineers*, 53–78.

15. Layton, *Revolt of the Engineers*, 55–59.

16. The historian Cecelia Tichi documents the trope in over one hundred silent films and in popular novels with sales of five million copies between 1897 and 1920. Cecelia Tichi, *Shifting Gears: Technology, Literature, Culture in Modernist America* (Chapel Hill: University of North Carolina Press, 1987), 98–99. Few examples reveal the ideal engineer as the 1930 pageant *Control*, staged to celebrate the ASME's 50th anniversary. It was written and directed by Yale theater professor George Pierce Baker and performed by students at the Stevens Institute

of Technology. Weaving together lavish costumes, projected motion pictures, and electronic music, it recounted the history of humanity as the history of engineering. George Pierce Baker and ASME, *Control: A Pageant of Engineering Progress* (New York: The American Society of Mechanical Engineers, 1930). Bruce Sinclair, "Local History and National Culture: Notions on Engineering Professionalism in America," in *American Technology*, Carroll W. Pursell, ed. (Malden, MA: Blackwell, 2001), 145–54.

17. Noble, *America by Design*, 33–48.

18. Roland Marchand, *Creating the Corporate Soul: The Rise of Public Relations and Corporate Imagery in American Big Business* (Berkeley: University of California Press, 1998). James Oliver Robertson argues that this form of contradiction is pervasive in cultural myths: "Very often, the problem being 'solved' by a myth is a contradiction of a paradox, something which is beyond the power of reason or rational logic to resolve. But the telling of the story, or the re-creation of a vivid and familiar image which is part of a myth, carries with it—for those who are accustomed to the myth, those who believe it—a satisfying sense that the contradiction has been resolved, the elements of the paradox have been reconciled." James Oliver Robertson, *American Myth, American Reality* (New York: Hill and Wang, 1980), 6.

19. Layton, *Revolt of the Engineers*, 225–48. See also William E. Akin, *Technocracy and the American Dream: The Technocrat Movement, 1900–1941* (Berkeley: University of California Press, 1977).

20. Ralph E. Flanders, "Engineering, Economics, and the Problem of Social Well Being," *Mechanical Engineering* 53, no. 2 (February 1931): 99–104; C. F. Hirshfeld, "Whose Fault?" *Mechanical Engineering* 54, no. 3 (March 1932): 173–80; Charles Jay Seibert, "Principal or Accessory to the Crime?" *Mechanical Engineering* 54, no. 9 (September 1932): 613–17.

21. Meiksins, "Engineers in the United States," 77–81.

22. This judgment comes from reading every issue of *Mechanical Engineering* from 1919 to 1941.

23. Thomas P. Hughes, *American Genesis: A Century of Invention and Technological Enthusiasm, 1870–1970* (New York: Viking, 1989), 353–442.

24. Joshua Stoff, *Picture History of World War II American Aircraft Production* (New York: Dover, 1993), ix-xii. See also Air Policy Commission, "Survival in the Air Age," in *A Report to the President's Air Policy Commission* (Washington, DC: 1948), reprinted in Carroll W. Pursell, ed., *The Military-Industrial Complex* (New York: Harper and Row, 1972), 178–97.

25. Small War Plants Corporation, *Economic Concentration and World War II*, ed. US Senate, Special Committee to Study Problems of American Small Business, 79th Cong. 206 (1946), reprinted in Pursell, *Military-Industrial Complex*, 151–77.

26. W. W. Rostow, *The Stages of Economic Growth: A Non–Communist Manifesto* (Cambridge: Cambridge University Press, 1960), 79, 170.

27. Carrol W. Pursell, appendixes to *Military-Industrial Complex*, 322–24.

28. Corning, "Rocketing into Your Daily Life," *Life* 47, no. 13 (September 28, 1959), 84; Chrysler, "The One Car Maker Who Makes Missiles Comes up with a New Way to Make Cars," *Life* 47, no. 15 (October 12, 1959), 157. See also: Cynthia Lee Henthorn, *From Submarines to Suburbs: Selling a Better America, 1939–1959* (Athens, OH: Ohio University Press, 2006).

29. National Science Foundation, *Employment of Scientists and Engineers in the United States, 1950–1966* (Washington, DC: US Government Printing Office, 1968), 6.

30. National Science Foundation, *Geographic Distribution of Federal Funds for Research and Development: Fiscal Year 1965* (Washington, DC: US Government Printing Office, 1967). For historical analysis, see: Robert Kargon, Stuart W. Leslie, and Erica Schoenberger, "Far beyond Big Science: Science Regions and the Organization of Research and Development," in *Big Science: The Growth of Large-Scale Research*, Peter Galison and Bruce Hevly, eds. (Stanford: Stanford University Press, 1992), 334–54; Bruce J. Schulman, *From Cotton Belt to Sunbelt: Federal Policy, Economic Development, and the Transformation of the South, 1938–1980* (New York: Oxford University Press, 1991); Christophe Lécuyer, *Making Silicon Valley: Innovation and the Growth of High Tech, 1930–1970* (Cambridge: MIT Press, 2006).

31. Robert Perrucci and Joel Emery Gerstl, eds., "Introduction" to *The Engineers and the Social System* (New York: Wiley, 1969), 2; Jeffrey M. Schevitz, *The Weaponsmakers: Personal and Professional Crisis during the Vietnam War* (Cambridge, MA: Schenkman, 1979), 5–7.

32. About the only constants among American engineers were their sex and race—less than 1 percent of all engineers were women, and people of color accounted for little more. Stanley S. Robin, "The Female in Engineering," in *The Engineer and the Social System*, Robert Perrucci and Joel Gerstl, eds. (New York: Wiley, 1969), 203–18. Even after the civil rights and women's movements, participation of women and minorities in engineering remained low. In 1993 bachelor's degrees in engineering received by women accounted for less than 20 percent of the total degrees awarded. Pamela E. Mack, "What Difference Has Feminism Made to Engineering in the Twentieth Century?" in *Feminism in Twentieth-Century Science, Technology, and Medicine*, Angela N. H. Creager, Elizabeth Lunbeck, and Londa L. Schiebinger, eds., Women in Culture and Society Series (Chicago: University of Chicago Press, 2001), 149–68; Amy E. Slaton, *Race, Rigor, and Selectivity in U.S. Engineering* (Cambridge: Harvard University Press, 2010).

33. Engineers Joint Council, *A Profile of the Engineering Profession: A Report from the 1969 National Engineers Register* (New York: EJC, 1971), 3.

34. Stuart W. Leslie, *The Cold War and American Science: The Military-Industrial-Academic Complex at MIT and Stanford* (New York: Columbia University Press, 1993); Roger L. Geiger, *Research and Relevant Knowledge: American Research Universities since World War II* (New York: Oxford University Press, 1993).

35. Leslie, *Cold War and American Science*, 9.

36. Furnas and McCarthy, *The Engineer*, 86–99.

37. Gordon S. Brown, "The Engineering of Science," *Technology Review* 62, no. 2 (December 1959): 19–22, 48–49.

38. Thomas P. Hughes, *Rescuing Prometheus* (New York: Pantheon Books, 1998), 4, 8, 21.

39. Bruce E. Seely, "Research, Engineering, and Science in Engineering Colleges, 1900–1960," *Technology and Culture* 34, no. 2 (April 1993): 344–86.

40. Meiksins, "Engineers in the United States," 83. Lécuyer, *Making Silicon Valley*, 169–209.

41. Paul Herbert Norgren and Aaron W. Warner, *Obsolescence and Updating of Engineers' and Scientists' Skills: Final Revised Report* (New York: Columbia University Seminar on Technology and Social Change, and United States Office of Manpower Policy Evaluation and Research, 1966), 15–64; Harold G. Kaufman, ed., *Career Management: A Guide to Combating Obsolescence* (New York: IEEE Press, 1975).

42. US Bureau of Labor Statistics, *Scientists, Engineers, and Technicians in the 1960s: Requirements and Supply* (Washington, DC: US Government Printing Office, 1963); Engineering Manpower Commission, *Engineering Student Attrition: Is It Undermining Our Nation's Engineering Manpower?* (New York: EJC, 1963); especially alarming was MIT's Presidential Report of 1960, which indicated that a majority of incoming students declared themselves interested in science rather than engineering. "The Great Engineering Debate, or Whither Goest Engineering?" *Chemical Engineering Progress* 59, no. 8 (August 1963), 15–18.

43. Dian Olson Belanger, *Enabling American Innovation: Engineering and the National Science Foundation* (West Lafayette, IN: Purdue University Press, 1998), 24–27, 43, 56.

44. Engineers Joint Council, *National Engineering Problems: Summary Report* (New York: EJC, 1964b), i–iv.

45. J. Douglas Brown, "Your Learned Profession," *Mechanical Engineering* 85, no. 3 (April 1963): 42–43.

46. Arnold J. Gully, "The Engineer and Technician: Their Similarities and Differences," *Chemical Engineering Progress* 63, no. 5 (May 1967): 26–29; T. O. Nethery, "The Engineer and Technician: A Symbiotic Relationship," *Chemical Engineering Progress* 63, no. 5 (May 1967): 31–32; A. V. Willett Jr. "The Trend of the Future," *Chemical Engineering Progress* 63, no. 5 (May 1967): 41–43.

47. Engineers Joint Council, *Engineering Manpower in Profile* (New York: EJC, 1964a), 7–9. See also National Science Foundation, *Employment of Scientists and Engineers*, 13; National Science Foundation, *Scientific and Technical Manpower Resources: Summary Information on Employment, Characteristics, Supply, and Training* (Washington, DC: US Government Printing Office, 1964), 86.

48. Perrucci and Gerstl found that over 60 percent of physics undergraduates were employed in engineering jobs. Robert Perrucci and Joel Emery Gerstl, *Profession without Community: Engineers in American Society* (New York: Random House, 1969), 75.

49. Noble quipped that titles of "scienteer" and "engitist" ought to exist to describe the hybridity of cold war R&D. Daniel E. Noble, *Noble Comments* (Phoenix, AZ: Motorola, 1970), 76–78.

50. David Kaiser, "The Postwar Suburbanization of American Physics," *American Quarterly* 56, no. 4 (December 2004), 854.

51. C. S. Draper, J. H. Kennan, T. K. Sherwood, and J. B. Wilbur, "Engineering and Education: A Statement Prepared by a Committee of the School of Engineering, MIT, March, 1961," *Journal of Engineering Education* 51, no. 10 (June 1961): 800.

52. William H. Whyte Jr., *The Organization Man* (New York: Simon and Schuster, 1956), 88–104.

53. C. Wright Mills, *White Collar: The American Middle Classes* (New York: Oxford University Press, 1951), 156–60.

54. John Dustin Kemper, *The Engineer and His Profession* (New York: Holt, Rinehhart, and Winston, 1967), 7.

55. Friden, "Your Beard Won't Bug Us," *IEEE Spectrum* 3, no. 8 (August 1966): 165; American Oil Company, "Did You Major in Restrictive Engineering?" *Mechanical Engineering* 88, no. 7 (July 1966): 141; Dow, "Must a Big Company Be Impersonal? We Think Not," *Technology Review* 66 (February 1964): 1; Friden, "I Got Fed up with Engineering Anonymous," *Mechanical Engineering* 90, no. 5 (May 1968): 145; Kelsey-Hayes, "Do Not Fold, Crumple, Staple or Otherwise Mutilate!" *Technology Review* 68 (November 1966): 72.

56. Vannevar Bush, "Trends in Engineering," *Tech Engineering News* 47 (November 1965): 13–17.

57. Wachs, *Careers in Engineering*, 15, 37–39.

58. Alan E. Nourse, *So You Want to Be an Engineer* (New York: Harper and Row, 1962), 2.

59. Engineers Joint Council, *National Engineering Problems*, 24–25.

60. Most came directly from the editor of *Chemical Engineering Progress*. Larry Resen, "Federal Impact Grows and Grows," *Chemical Engineering Progress* 60, no. 2 (February 1964): 22–24; Larry Resen, "Government, Government, Government," *Chemical Engineering Progress* 60, no. 2 (February 1964): 31; Larry Resen, "The Handwriting on the Blackboard," *Chemical Engineering Progress* 60, no. 5 (May 1964): 43; Nels E. Sylvander, "The Myth of Government/Industry Partnership," *Chemical Engineering Progress* 61, no. 9 (September 1965): 25–27.

61. W. R. Marshall Jr., "Science Ain't Everything," *Chemical Engineering Progress* 60, no. 1 (January 1964): 17–21.

62. C. C. Furnas, "Engineering Is Not Obsolescent," *Chemical Engineering Progress* 59, no. 8 (August 1963): 20–21; J. F. Skelly, "Coexistence Needed in Education," *Chemical Engineering Progress* 59, no. 8 (August 1963): 22–23.

63. Larry Resen, "Treating Professionals Professionally," *Chemical Engineering Progress* 58, no. 9 (September 1962), 35; Meiksins, "Engineers in the United States," 91; Schevitz, 17–31. Perrucci and Gerstl found that only 11 percent of engineers rated belonging to a professional community outside of work as "very important" and over half identified immediate superiors as those best suited to judge performance. PhD engineers had a higher sense of professionalism, but even they identified their bosses as better judges of professional performance over fellow engineers. Perrucci and Gerstl, *Profession without Community*, 103, 117.

64. Kemper, *Engineer and His Profession*, iii.

65. Ibid., iii.

66. J. William Fulbright, "The Great Society Is a Sick Society," *New York Times Magazine* (August 20, 1967): 30, 88–96; Perlstein, *Nixonland*, 373–96.

67. "The Moon and 'Middle America,'" *Time* (August 1, 1969): 10–11.

68. Dael Wolfle, "The Big Story—The Meaning of Science and Technology in Modern Society. Do the Engineering Society Publications Cover It?" in *Engineering Societies and Their Literature Programs*, ed. Larry Resen (New York: Engineers Joint Council, 1967), 63–66.

69. Westinghouse Electric Corparation, "Go Westinghouse, Young Man!" *Engineering and Science* 30, no. 4 (January 1967): 1.

70. Ford Motor Company, "We're One of the Causes of Air Pollution," *Technology Review* 73, no. 8 (June 1971): 18; Pacific Telephone, "Truce," *Daily Titan* (January 12, 1971): 4.

71. Westinghouse, "Who Needs Engineers?" *Engineering Education* 62, no. 3 (December 1972): 161.

72. William W. Hill, "What Is Industry Doing about Poverty and Prejudice?" *Professional Engineer* 38, no. 10 (October 1968): 27–31. See also J. T. Kane, "Dialogue with Some Very Young Engineers," *Professional Engineer* 39, no. 11 (November 1969): 30–34.

73. Stanley W. Burriss, "An Industry View," in *Women in Engineering: Bridging the Gap between Society and Technology*, George Bugliarello, Vivian Cardwell, Olive Salembier, and Winifred White, eds. (Chicago: University of Illinois at Chicago Circle, 1971), 15.

74. M.D. "From the Editor's Notebook," *Tech Engineering News* 46 (March 1965): 9.

75. Jerome Lettvin, "You Can't Even Step in the Same River Once," *Tech Engineering News* (December 1967): 35–41.

76. Says Beers: "My father started carrying these thick books with him to and from work. He would read at Lockheed on his lunch hour and in the living room after dinner. He became a student of John Kenneth Galbraith's critique of corporate life as the misplaced pursuit of money over life's other, more fulfilling rewards. He soaked up Eric Hoffer's idea that mass movements are the products

of adults gripped by 'juvenile' restlessness, which, in turn, is produced by economic dislocation due to technology. . . . He read a number of books about race and prejudice, including The *Autobiography of Malcolm X* and Eldridge Cleaver's *Soul on Ice*. He took to heart the portrait of a shrinking planet offered by Paul Ehrlich and Alvin Toffler. My father's titles ran into one another like haikus of gloom." David Beers, *Blue Sky Dream: A Memoir of America's Fall from Grace* (New York: Doubleday, 1996), 121–33. This is in marked contrast to another recent memoir of an engineer's offspring. M. G. Lord remembers that her father's nonengineering books included John A. Stormer's *None Dare Call It Treason*, J. Edgar Hoover's *Masters of Deceit: The Story of Communism in America*, and Phyllis Schlafly's *A Choice, Not an Echo*. M. G. Lord, *Astro Turf: The Private Life of Rocket Science* (New York: Walker, 2005), 60.

77. National Science Foundation, *Unemployment Rates and Employment Characteristics for Scientists and Engineers, 1971* (Washington, DC: US Government Printing Office, 1972), 4, 79.

78. Paula Goldman Leventman, *Professionals out of Work* (New York: Free Press, 1981), 46–47.

79. Robert Hotz, "SST Postmortem," *Aviation Week and Space Technology* 95 (April 5, 1971): 9.

80. Don McAllister, "Letter to the Editor," *Aviation Week and Space Technology* 95 (April 5, 1971), 50.

81. Henri Dupre, "Letter to the Editor," *Aviation Week and Space Technology* 95 (April 5, 1971): 50; John L. Pedrick Jr., "Letter to the Editor: National Priorities," *Aviation Week and Space Technology* 95 (April 12, 1971): 60.

82. Anonymous, *Mute Engineers*, 18–48.

83. Ibid., 17.

84. Ibid., 5, 10.

85. Sandra B. Goldsmith, "Engineering under Attack: An NSPE Meeting Panel," *Professional Engineer* 40, no. 11 (March 1970): 38–39.

86. Arthur Kantrowitz, "The Test," *Tech Engineering News* 50, no. 7 (December 1968): 12–15.

87. George M. Newcombe, "Engineering, a Modern Profession in a Moral Society—A Student's Viewpoint," in *Are Engineering and Science Relevant to Moral Issues in a Technological Society?* (New York: EJC, 1969), 7–11.

Chapter 3

1. Ellul, *Technological Society*, 78.

2. Arthur M. Schlesinger Jr., "The Velocity of History," *Newsweek* (July 6, 1970): 32–34.

3. Daniel Bell, *The End of Ideology: On the Exhaustion of Political Ideas in the Fifties* (Glencoe, IL: Free Press, 1960), 36.

4. Students for a Democratic Society, *The Port Huron Statement* (New York: Students for a Democratic Society, 1962), 22.

5. Mumford, *Technics and Human Development*, 3.

6. Federal funding of the social sciences increased sixfold in the 1960s. Walter McDougall, *The Heavens and the Earth: A Political History of the Space Age* (New York: Basic Books, 1985), 440–43. On the paperback press see Marshall A. Best, "In Books, They Call It a Revolution," *Daedalus* 92 no. 1 (winter 1963), 30–41.

7. Winner, *Autonomous Technology*, 55.

8. Thomas P. Hughes, introduction to *Changing Attitudes toward American Technology*, Thomas P. Hughes, ed. (New York: Harper and Row, 1975), 3.

9. Lawrence S. Wittner, *Resisting the Bomb: A History of the World Nuclear Disarmament Movement, 1945–1970* (Stanford, CA: Stanford University Press, 1997).

10. Rachel Carson, *Silent Spring* (Boston: Houghton Mifflin, 1962); Michael Egan, *Barry Commoner and the Science of Survival: The Remaking of American Environmentalism* (Cambridge: MIT Press, 2007).

11. Ralph Nader, *Unsafe at Any Speed: The Designed-in Dangers of the American Automobile* (New York: Grossman, 1965).

12. Martin Luther King Jr., *Strength to Love* (New York: Harper and Row, 1963), 57.

13. Charles A. Reich, *The Greening of America* (New York: Random House, 1970).

14. Winner, *Autonomous Technology*, 11.

15. Ibid., 46.

16. Ibid., 178–87.

17. Ibid., 208–26.

18. Leo Marx, "The Idea of 'Technology' and Postmodern Pessimism," in Smith and Marx, *Does Technology Drive History?*, 238–57.

19. Ellul, *Technological Society*, xxv.

20. Mumford, *Technics and Human Development*, 234–42.

21. Marcuse, *One Dimensional Man*, xv–xvi.

22. Lewis Mumford, *The Myth of the Machine: The Pentagon of Power* (New York: Harcourt Brace Jovanovich, 1970), 334, 420.

23. Marcuse, *One Dimensional Man*, 220–24.

24. John Kenneth Galbraith, *The New Industrial State* (Boston: Houghton Mifflin, 1967), 282–95, 370–78.

25. Reich, *Greening of America*, 358.

26. John M. Boyd, "Science Is Dead—Long Live Technology!" *Engineering Education* 62, no. 8 (1972): 892–95.

27. J. D. Horgan, "Technology and Human Values: The 'Circle of Action,'" *Mechanical Engineering* 95, no. 8 (August 1973): 19–22.

28. "Evolution of the BBN Underground," *Signal/Noise* 1, no. 1 (1970), 1.

29. See, for example, "Unemployment of Scientists and Engineers," *GE Resistor* 3 (January–February 1971): 1–2.

30. Alvin W. Gouldner, *The Dialectic of Ideology and Technology: The Origins, Grammar, and Future of Ideology* (New York: Seabury Press, 1976), 23–66.

31. William Fielding Ogburn, *Social Change with Respect to Culture and Original Nature* (New York: Huebsch, 1922), 201. For an analysis of Ogburn's ideas see David McGee, "Making up Mind: The Early Sociology of Invention," *Technology and Culture* 36, no. 4 (October 1995): 773–801.

32. William Fielding Ogburn, *You and Machines* (Chicago: University of Chicago, 1934), 1.

33. United States President's Research Committee on Social Trends, *Recent Social Trends in the United States: Report of the President's Research Committee on Social Trends* (New York: McGraw-Hill, 1933), passim. Culture lag then became the "law of unequal rates of change." William Fielding Ogburn, "Laggard Parts of Our Social Machine," *New York Times Magazine* (April 16, 1933): 5, 19. Similarly in a *Journal of Business* article from 1936 he recycled his argument from *Social Change* with "technological change" and "technology" replacing "invention." William Fielding Ogburn, "Technology and Governmental Change," *Journal of Business of the University of Chicago* 9, no. 1 (1936): 1–13.

34. Peter Drucker, *Landmarks of Tomorrow* (New York: Harper, 1959); Kenneth Ewart Boulding, *The Meaning of the Twentieth Century: The Great Transition* (New York: Harper and Row, 1964); Robert L. Heilbroner, *The Future as History: The Historic Currents of Our Time and the Direction in Which They Are Taking America* (New York: Harper, 1960); Margaret Mead, "Preface, " in *Cultural Patterns and Technical Change* (New York: New American Library and United Nations Educational, Scientific and Cultural Organization, 1958): 5–7.

35. Barbara Ward, "We Are All Developing Nations," *New York Times Magazine* (February 25, 1962): 4, 38, 40, 43. Barbara Ward, *Spaceship Earth* (New York: Columbia University Press, 1966).

36. Interim-Committee on the Social Aspects of Science, "Society in the Scientific Revolution," *Science* 124, no. 3234 (December 21, 1956): 1231. Price made the case that the "high-velocity change" of the unfolding "Scientific Revolution" upset government's checks and balances, requiring new collaborations between scientists and policymakers. Don K. Price, *The Scientific Estate* (Cambridge: Belknap Press of Harvard University Press, 1965), 214, 278.

37. Bix, *Inventing Ourselves out of Jobs,* 236–79.

38. Norbert Wiener, *The Human Use of Human Beings: Cybernetics and Society* (Boston: Houghton Mifflin, 1954). See also Robert K. Merton, "The Machine,

the Worker, and the Engineer," *Science* 105, no. 2717 (January 24, 1947): 79–84.

39. John Diebold, *Automation: The Advent of the Automatic Factory* (New York: Van Nostrand, 1952), 148–75.

40. Marshall McLuhan, *Understanding Media* (New York: McGraw-Hill, 1964), 21.

41. Howard Brick, "Optimism of the Mind: Imagining Post-industrial Society in the 1960s and 1970s," *American Quarterly* 44 (September 1992): 348–80.

42. Harvard University Program on Technology and Society, *First Annual Report of the Executive Director* (Cambridge: Harvard University Program on Technology and Society, 1965), 1. Insight into IBM's motivations comes from David Drew of the Claremont Graduate University, who was a software engineer at the Harvard University Computing Center in the early 1960s. David Eli Drew, "Why Don't All Professors Use Computers?" *Academic Computing* 4, no. 2 (October 1989): 12–14, 58–60.

43. "Technological Change: Processes, Impacts, and Adjustment Policies," 25 March 1964, box 1, folder "Birth of the Program," Records of the Harvard University Program on Technology and Society, UAV.825.16, Harvard University Archives [hereafter cited as Harvard Program Records].

44. Mesthene was a jack-of-all-trades. During World War II he helped create a Chinese–English dictionary so that American soldiers could communicate with the Chinese resistance. He then assisted in de-Nazification programs for German POWs in New Jersey. After the war, he received his BA in philosophy from Columbia. He briefly taught at Adelphi College, before spending two years as an editor for Bantam Books. While at RAND, he serving as the conductor of the company orchestra and worked toward his doctorate in philosophy with a dissertation titled "How Language Makes Us Know." In the two years prior to his appointment at Harvard, he was the Secretary of the Ministerial Meeting on Science for the Organization of Economic Co-operation and Development (OECD) in Paris. Emmanuel G. Mesthene and Organisation for Economic Cooperation and Development, *Ministers Talk About Science: A Summary and Review* (Paris: Organisation for Economic Co-operation and Development, 1965). In his RAND projects Mesthene employed a methodology similar to Ogburn's invention studies, analyzing the government's attempt to develop titanium as an aerospace material. Emmanuel G. Mesthene, *The Titanium Decade* (Santa Monica, CA: RAND Corporation, 1962).

45. Harvard Program, *First Annual Report*, 7–8.

46. Harvard University Program on Technology and Society, *Third Annual Report of the Executive Director* (Cambridge: Harvard University Program on Technology and Society, 1967), 15.

47. Harvey Brooks to George Baker, Don Price, and Emmanuel G. Mesthene, memorandum, 2 May 1968, box 1, folder "EGM/IBM/Horton," UAV.825.12, Harvard Program Records.

48. William K. Stevens, "Study Terms Technology a Boon to Individualism," *New York Times* (January 18, 1969): 1, 17; "A Student of Technology: Emmanuel George Mesthene," *New York Times* (January 18, 1969): 17.

49. Emmanuel G. Mesthene, "Some General Implications of the Research of the Harvard University Program on Technology and Society," *Technology and Culture* 10, no. 4 (1969a), 489–92.

50. Emmanuel G. Mesthene, "How Technology Will Shape the Future," *Science* 161, no. 3837 (1968): 135–43.

51. For a complete list of publications see: Harvard University Program on Technology and Society, *A Final Review, 1964–1972* (Cambridge: Harvard University Press, 1972), 245–64.

52. "External Relations of the Program 1967–1970," box 1, UAV825.18, Harvard Program Records.

53. The Program boasted a readership of four hundred members of the federal government. Policy had always been part of Mesthene's agenda. He questioned the efficiency of having scientists make science policy and claimed that new scientifically literate policy makers were better suited to plan the nation's sociotechnical future. Emmanuel G. Mesthene, "Can Only Scientists Make Government Science Policy?" *Science* 145, no. 3629 (July 17, 1964): 237–40; Emmanuel G. Mesthene, "The Impacts of Science on Public Policy," *Public Administration Review* 27, no. 2 (1967a): 97–104.

54. Committee on Science and Astronautics US House of Representatives, *Technology: Processes of Assessment and Choice, Report of the National Academy of Sciences* (Washington, DC: US Government Printing Office, 1969), 1–12, 137. See also Sylvia Doughty Fries, "Expertise against Politics: Technology as Ideology on Capitol Hill, 1966–1972," *Science, Technology, and Human Values* 8, no. 2 (spring 1983): 6–15; Bruce A. Bimber, *The Politics of Expertise in Congress: The Rise and Fall of the Office of Technology Assessment* (Albany: State University of New York Press, 1996).

55. *Technology Assessment: Hearings before the Subcommittee on Science, Research, and Development of the Committee on Science and Astronautics, U.S. House of Representatives*, 91st Cong., 13 (1969b) (statement of Dr. Mesthene, Director, Program on Technology and Society, Harvard University).

56. Simon Ramo, *The Business of Science: Winning and Losing in the High-Tech Age* (New York: Hill and Wang, 1988), 32–35.

57. Hughes, *Rescuing Prometheus*, 69–140.

58. Simon Ramo, "Weapons Systems Engineering and Changing Role of the Scientist," *Armed Forces Chemical Journal* (July–August 1956): 23; Simon Ramo, "The Impact of Systems Engineering on Education," *American Engineer* 27, no. 10 (October 1957): 11–15; Simon Ramo, keynote address Case Systems Symposium, 26 April 1960, box 53, folder 14, Simon Ramo Papers, University of Utah Library [hereafter cited as Ramo Papers]; Simon Ramo, "The Impact of Missiles and Space on Electronics," *Proceedings of the Institute of Radio Engineers* 50, no. 5 (May 1962): 1237–41.

59. "The Coming Technological Society," *Computers and Automation* 10, no. 7 (July 1961): 15–16, 18–21 reprinted in Simon Ramo, "The Coming Technological Society," *The Executive* 5, no. 5 (October 1961): 11–13. James Winchester repeated Ramo's text almost verbatim in James Winchester, "Firm Looks Ahead in Technology," *Christian Science Monitor* (December 5, 1962): 1. An augmented version appeared in Simon Ramo, "The Coming Technological Society," *NATO's Fifteen Nations* (December 1964–January 1965): 66–73, which made explicit the dominance of technology over all other factors of social change.

60. Ronald Reagan, "Science for Mankind," 8 November 1967, box 10, folder 11, Ramo Papers; Ronald Reagan to Simon Ramo, 2 November 1967, box 10, folder 11, Ramo Papers.

61. Simon Ramo, *Cure for Chaos: Fresh Solutions to Social Problems through the Systems Approach* (New York: McKay, 1969b).

62. Ramo, *Century of Mismatch*, 9–10.

63. Ibid., 85.

64. Ibid., 116–26.

65. Ibid., 163–85.

66. The only hint of Ramo's intellectual formation—which suggests he was influenced by Ogburn—was an oft deployed anecdote about participation in a college debate about the engineers' responsibility for the Great Depression. Ramo, *Century of Mismatch*, vi; Simon Ramo, "The Coming Shortage of Educated People," *Engineering Education* 63, no.1 (October 1972b): 19–20.

67. Ramo, *Century of Mismatch*, vii, 11–12.

68. "The Misuse of Science," *The Nation* (February 24, 1969): 228.

69. James J. Harford, "AIAA: Meet the Social Scientist," *Astronautics and Aeronautics* 8, no. 8 (August 1970): 19–21.

70. Richard C. Dorf, "Technology and Man's Future: University of Santa Clara April 1968," *Technology and Culture* 9, no. 4 (October 1968): 580–84.

71. Bruce E. Seely, "SHOT, the History of Technology, and Engineering Education," *Technology and Culture* 36, no. 4 (October 1995): 739–72.

72. Simon Ramo, "Comment: The Anticipation of Change," *Technology and Culture* 10, no. 4 (October 1969a): 514–21.

73. Eric A. Walker, "Engineers: A Time for Leadership," in *Engineering, Technology and Society: the Gwilym A. Price Engineering Lectures*, H. E. Hoelscher, ed. (Pittsburgh: University of Pittsburgh, 1972), 41.

74. Ibid., 40.

75. National Academy of Engineering, *A Study of Technology Assessment: Report of the Committee on Public Engineering Policy* (Washington, DC: US Government Printing Office, 1969), vii.

76. Ibid., 20.

77. John R. Moore, "Changes in Management and the Management of Change," *IEEE Transactions on Aerospace and Electronic Systems* AES-7, no. 5 (November 1969): 1024–27.

78. Walter H. Kohl, "Assessment, Transfer, and Forecasting of Technology," *IEEE Spectrum* 8, no. 1 (January 1971): 70–75.

79. David W. Ewing, ed. *Technological Change and Management: The John Diebold Lectures, 1968–1970* (Boston: Harvard University Graduate School of Business Administration, 1970).

80. Harvey Brooks, "Dilemmas of Engineering Education," *IEEE Spectrum* 4, no. 2 (February 1967): 89–91.

81. H. K. Nason, "Management's Changing Responsibilities," *Mechanical Engineering* 81, no. 10 (October 1959): 42–44.

82. Allen F. Rhodes, "Evolution and Technology in Conflict," *Mechanical Engineering* 93, no. 3 (March 1971): 18–19.

83. John McDermott, "Technology: The Opiate of the Intellectuals," *The New York Review of Books* (July 31, 1969): 25–36.

84. Emmanuel G. Mesthene, "Technology and Wisdom," in *Technology and Social Change*, ed. Emmanuel G. Mesthene (Indianapolis: Bobbs-Merrill, 1967b), 61.

85. Irene Taviss and Linda Silverman, *Technology and Values* (Cambridge: Harvard University Program on Technology and Society, 1969).

86. Irene Taviss, "A Survey of Popular Attitudes toward Technology," *Technology and Culture* 13, no. 4 (October 1972): 606–21.

87. George Basalla, "Addressing a Central Problem," *Science* 180, no. 4086 (May 11, 1973), 584.

88. Mumford, *Pentagon of Power*, 208.

89. Winner, *Autonomous Technology*, 319.

90. This merger of positions usually was the result of unfamiliarity or of deliberate poaching by engineer readers, but it sometimes was the consequence of how intellectuals and publishers framed texts. In his foreword to the English translation of the *Technological Society*, the sociologist Robert K. Merton—a Harvard Program advisor—summarized what he took to be the book's message: "The essential point, according to Ellul, is that technique produces all this without plan. . . . Our technical civilization does not result from a Machiavellian scheme. It is a response to the 'laws of development' of technique." Robert K. Merton, foreword to Ellul, *Technological Society*, viii.

91. Roy Rulseh, "To the Editor," *Mechanical Engineering* 92, no. 3 (March 1970): 58.

92. Samuel S. Baxter, "Blueprint for the New ASCE Year," *Civil Engineering* 42, no. 1 (January 1971): 43–45.

Chapter 4

1. ASME Technology & Society Division, "Proposal: Fundamental Principles of Engineering Ethics," 26 April 1973, box 7, ASME: Ethics Committee, Paschkis Papers.

2. C. Wright Mills, *The Power Elite* (New York: Oxford University Press, 1956), 20–27, 343–61.

3. Oscar S. Bray, "ASCE and Social Issues," *Civil Engineering* 41, no. 5 (May 1971): 58–59.

4. Harry C. Simrall, "If You Are a Concerned Engineer . . . ." *IEEE Spectrum* 8, no. 2 (February 1971): 69–71.

5. Layton, *Revolt of the Engineers*, 171.

6. W. L. Abbott, "To New ASME Members," *Mechanical Engineering* 52, no. 1 (January 1930): front cover, 1.

7. D. Freiday, "The Vanishing Engineer?" *Mechanical Engineering* 82, no. 7 (July 1960): 53–54; Philip Sporn, "The Case for the Engineer," *Mechanical Engineering* 83, no. 6 (June 1961): 42–44; Martin Goland, "Engineering Geriatrics," *Mechanical Engineering* 84, no. 1 (January 1962): 26–27.

8. R. A. Sherman, "The ASME *IS* a Professional Society," *Mechanical Engineering* 84, no. 2 (February 1962): 30–31.

9. Ronald B. Smith, "Professional Responsibility of Engineering," *Mechanical Engineering* 86, no. 1 (January 1964): 18–20.

10. G. M. Gilbert, *Nuremberg Diary* (New York: Farrar, Strauss, and Giroux, 1947); Hannah Arendt, *Eichmann in Jerusalem: A Report on the Banality of Evil* (New York: Viking, 1963).

11. Alice Kimball Smith, *A Peril and a Hope: the Scientists' Movement in America, 1945–47* (Cambridge: MIT Press, 1970); Jessica Wang, *American Science in an Age of Anxiety: Scientists, Anticommunism, and the Cold War* (Chapel Hill: University of North Carolina Press, 1999); S. S. Schweber, *In the Shadow of the Bomb: Bethe, Oppenheimer, and the Moral Responsibility of the Scientist* (Princeton: Princeton University Press, 2000).

12. Jeffrey C. Isaac, *Arendt, Camus, and Modern Rebellion* (New Haven: Yale University Press, 1992); H. Richard Niebuhr, *The Responsible Self: An Essay in Christian Moral Philosophy* (New York: Harper and Row, 1963); Dwight Macdonald, *The Responsibility of Peoples: and Other Essays in Political Criticism* (London: Gollancz, 1957); James M. Gustafson and James T. Laney, eds., *On Being Responsible: Issues in Personal Ethics* (New York: Harper and Row, 1968). William Horosz, *The Crisis of Responsibility: Man as the Source of Accountability* (Norman: University of Oklahoma Press, 1975); Chad Lavin, *The Politics of Responsibility* (Champaign-Urbana: University of Illinois Press, 2008).

13. Scientists scored a lowly forty points, earning the highest ranking in specialized training, but zero points in all other categories. Joseph G. Wilson, "High in Accomplishment, Yes; but Professional?" *Mechanical Engineering* 86, no. 6 (June 1964): 18–20.

14. Victor Paschkis, "To the Editor," *Mechanical Engineering* 86, no. 8 (August 1964): 74.

15. John Ryker, "To the Editor," *Mechanical Engineering* 86, no. 10 (October 1964): 77.

16. W. E. Little, "To the Editor," *Mechanical Engineering* 86, no. 12 (December 1964): 78.

17. F. E. Burke, "To the Editor," *Mechanical Engineering* 87, no. 2 (February 1965): 78.

18. James D. Thackrey, "To the Editor," *Mechanical Engineering* 88, no. 5 (May 1966): 159.

19. N. A. Christensen, "The Purpose of Professional Engineering Ethics," *Mechanical Engineering* 87, no. 11 (November 1965): 46–48. See also Philip L. Alger, N. A. Christensen, and Sterling P. Olmsted, *Ethical Problems in Engineering* (New York: Wiley, 1965).

20. Richard G. Folsom, "Technology and Humanism," *Mechanical Engineering* 88, no. 1 (January 1966): 20–23.

21. J. J. Jaklitsch Jr. "In Favor of Achievement," *Mechanical Engineering* 89, no. 12 (December 1967): 13.

22. J. Duffy, "To the Editor," *Mechanical Engineering* 89, no. 9 (September 1967): 92–93; Robert E. McKechnie "To the Editor" *Mechanical Engineering* 89, no. 10 (October 1967): 82; Thomas B. Speer, "To the Editor," *Mechanical Engineering* 89, no. 10 (October 1967): 83.

23. Bethlehem Steel, "Concern," *Tech Engineering News* 52, no. 7 (December 1970): 20. See also: GE, "On Your Way up in Engineering, Please Take the World with You," *Technology Review* 73, no. 2 (December 1970): 38; The Dow Chemical Company, "Which Are You . . . A Good-Doer or a Do-Gooder?" *Engineering Education* 60, no. 7 (March 1970): 707; Xerox, "For Engineers Who Think of More Than Engineering," *Mechanical Engineering* 92, no. 4 (April 1970): 131.

24. David Packard, "Management's Expanding Responsibilities," *Mechanical Engineering* 87, no. 6 (June 1965): 18–21.

25. Donald C. Burnham, "Productivity: Key to Progress," *Mechanical Engineering* 90, no. 9 (September 1968): 30–33.

26. Chauncey Starr, "Social Benefit versus Technological Risk," *Science* 165, no. 3899 (September 19, 1969): 1232–38.

27. Donald Marlowe, "The New Luddites," *Mechanical Engineering* 92, no. 3 (March 1970a): 12–13.

28. Victor Paschkis, "To the Editor," *Mechanical Engineering* 92, no. 6 (June 1970a): 81.

29. Stein Weissenberger, "To the Editor," *Mechanical Engineering* 92, no. 6 (June 1970): 82. See also Thomas Cherbas, "To the Editor," *Mechanical Engineering* 92, no. 5 (May 1970): 80.

30. Donald Marlowe, "As the President Sees It," *Mechanical Engineering* 91, no. 9 (September 1969c): 81.

31. Victor Paschkis "Autobiography," box 4, Rough Drafts (and Diaries) from Autobiography, Paschkis Papers.

32. Victor Paschkis "Chapter 5: Married," box 4, Rough Drafts (and Diaries) from Autobiography, Paschkis Papers.

33. Victor Paschkis, "The Heat and Mass Flow Analyzer Laboratory," *Metal Progress* 52, no. 6 (November 1947b): 813–18.

34. Victor Paschkis, "Notes for Chapter IX," box 4, Rough Drafts (and Diaries) from Autobiography, Paschkis Papers.

35. Victor Paschkis, "The Disease They-Mindedness and Its Cure," *Journal of Human Relations* 13, no. 2 (1965): 178–84; Victor Paschkis, "Double Standards," *Friends Intelligencer* (August 30, 1947a): 463.

36. Victor Paschkis, "Constituting Assembly" symposium flyer, 7 July 1949, Box 2, Race Relations: Report on Travels to Promote, Paschkis Papers.

37. Moore, *Disrupting Science*, 54–95.

38. Ad hoc Committee on the Triple Revolution. "The Triple Revolution," *Liberation* 9, no. 2 (April 1964): 9–15.

39. Victor Paschkis, "Cybernation and Civil Rights," in *The Evolving Society: The Proceedings of the First Annual Conference on the Cybercultural Revolution--Cybernetics and Automation*, ed. Alice Mary Hilton and Institute for Cybercultural Research (New York: Institute for Cybercultural Research, 1966), 357–64. The participants, many of them authors of the *Triple Revolution*, were philosophers, activists, and DOD executives ranging from Hannah Arendt to Paul Armer, head of the Computing Sciences Division at the RAND Corporation.

40. Victor Paschkis, "C.U. Institute for the Study of Scientific Affairs," 9 October 1968, box 1, Correspondence and Writings, Paschkis Papers.

41. D. N. Beshers, J. E. Englund, E. L. Gaden Jr., P. J. Kolesar, L. J. Lidofsky, S. Melman, V. Paschkis, M. G. Salvadori, W. T. Sanders, T. F. Stern, and S. H. Unger, "Proposal for Undergraduate Course," 12 January 1970, box 1, Correspondence and Writings, Paschkis Papers.

42. Victor Paschkis, "Technology and Society," 9 October 1968, p. II-7, box 4, Survival Manuscript, Paschkis Papers.

43. This was especially true of its leadership, with aerospace engineer T. Paul Torda of IIT and John T. Berry of the University of Vermont playing formative roles.

44. "Objectives of Committee on Technology and Society" 8 June 1970, box 3, Goals, Paschkis Papers.

45. Jesse Mock, ed., *The Engineer's Responsibility to Society* (New York: ASME, 1969).

46. Victor Paschkis, "Moving toward Responsible Technology," ASME Winter Annual Meeting, 71-WA/Av-1.

47. Victor Paschkis, "Assessment—By Whom, for Whom?" ASME Winter Annual Meeting, 70-WA/Av-5.

48. ASME Technology & Society Division, "Proposal: Fundamental Principles of Engineering Ethics," 26 April 1973, box 7, ASME: Ethics Committee, Paschkis Papers.

49. "Goals Report Committee, "Goals: A Proposed Statement," *Mechanical Engineering* 92, no. 4 (April 1970), 18.

50. Ibid., 19, 22.

51. Donald E. Marlowe, "The Specter of Socialized Engineering," *Mechanical Engineering* 86, no. 6 (June 1964b): 24–25; Donald E. Marlowe, "The Legacy of Merlin," *Mechanical Engineering* 86, no. 2 (February 1964a): 26–29.

52. Donald E. Marlowe, "Prometheus Unbound," *Mechanical Engineering* 91, no. 11 (November 1969d): 28–30.

53. Donald E. Marlowe, "Technology and Society, Part 1: The Public Interest," *Mechanical Engineering* 91, no. 4 (April 1969b): 24–26.

54. Goals Report Committee, "Goals: A Proposed Statement," 23; Donald E. Marlowe, "Goals for a More Vital Society," *Mechanical Engineering* 92, no. 4 (April 1970b): 17.

55. "Goals: Basis for Action Programs, 'Making Technology a True Servant of Man,'" *Mechanical Engineering* 93, no. 4 (April 1971): 16–19. Three goals proved more controversial than the rest. Goal 4, "economic needs," raised fears of unionization and to assure its passage the word "responsibility" in the preliminary version was replaced by "concern." Goal 7, "membership," which resulted in the acceptance into the ASME of new engineering-technologists with four-year BA degrees, raised longstanding disagreements about the exclusivity of the member societies; it was adopted with only 54 percent approval. Goal 14, "equal opportunities in engineering," which was not included in the preliminary statement, was the only Goal to receive less than a 50 percent majority, with 22 percent abstaining, and 30 percent opposed. "ASME Goals: Summary of Member Reaction Recorded at Discussion Sessions," 20 November 1970, box 6, Paschkis Papers.

56. Colin Carmichael, "The Role of ASME in Government," in *ASME Goals: A Collection of Papers Prepared for the ASME Goals Conference* (New York: ASME, 1970): H1–13.

57. Kenneth A. Roe, "President's Message," in *The ASME 1971–1972 Council Report to the Membership* (New York: ASME, 1972), 1.

58. Indeed Marlowe and others asserted that the Goals were not ethical canons, but rather "stars" by which to set the ASME's course. Goals Report Committee, "Goals: a Proposed Statement," 18.

59. James M. Singleton, Alan D. Anderson, Ronald Wayne Burr, Jack Gammill Clemens, K. Fred Rist, and Deborah M. Schmitz., "The Engineer and Society: Students Speak Out," *Mechanical Engineering* 93, no. 9 (September 1971): 33–35; "Goals the Pursuit of Goals One: An Appeal for Help," *Mechanical Engineering* 93, no. 9 (September 1971): 87.

60. J. J. Jaklitsch Jr. "ASME Issues Energy Policy Statement," *Mechanical Engineering* 98, no. 1 (January 1976), 3. The ASME also opened the pages of *Mechanical Engineering* to Ralph Nader, who offered a critique of nuclear

power; however, the ASME's Nuclear Power Codes and Standards Committee appended a fourteen-point dismissal to his article. Ralph Nader, "Nuclear Power: More than a Technical Issue," *Mechanical Engineering* 98, no. 2 (February 1976): 32–36; Committee on Nuclear Power, "ASME Postscript," *Mechanical Engineering* 98, no. 2 (February 1976): 36–37.

61. Committee on Technology and Society, "Suggested Modifications of the ASME Goals," 21 September 1970, box 8, ASME Goals, Paschkis Papers.

62. ABC, jr, "Role of the Policy Board General Engineering Department," 18 January 1974, box 7, ASME Defense, Paschkis Papers.

63. T. Paul Torda, "Position Paper: Technology and Society Division," 28 May 1975, box 7, ASME Defense, Paschkis Papers.

64. Victor Paschkis to Theodore Roszak, 3 July 1979, box 7, ASME Correspondence, Paschkis Papers.

65. William H. Wisely, "The Engineering Profession as It Really Is!" in *ASME Goals: A Collection of Papers Prepared for the ASME Goals Conference* (New York: ASME, 1970): C23–24.

66. Terry S. Reynolds, *75 Years of Progress: A History of the American Institute of Chemical Engineers, 1908–1983* (New York: AIChE, 1983), 101.

67. "What EEs Want," *IEEE Spectrum* 9 no. 3 (March 1972): 48–52. See also A. Michal McMahon, *The Making of a Profession: A Century of Electrical Engineering in America* (New York: IEEE, 1984), 259–63.

68. Leo Spector, "To the Editor," *Mechanical Engineering* 92, no. 7 (July 1970): 87.

69. Dennis Warner, "The Challenge to Civil Engineering," *Civil Engineering* 42, no. 7 (July 1972): 66–67.

70. Donald E. Marlowe, "Public Interest—First Priority in Engineering Design?" *Professional Engineer* 39, no. 2 (February 1969a): 23–25.

71. In recognizing unintended effects as a serious challenge to traditional professionalism, he prefigured sociologist Ulrich Beck's *Risk Society*. As Beck expressed it, restoring expert credibility "lies in the responsibility for side effects itself" because the processes of defining risks "create zones of *illegitimate* systemic conditions, which cry out for change in the interest of the general public." Ulrich Beck, *Risk Society: Toward a New Modernity* (London: Sage Publications, 1992), 227.

72. Donald Christiansen, ed. "Future Shock for EEs," *IEEE Spectrum* 9, no. 1 (January 1972): 39.

73. Hamilton Standard, "Wanted: An Engineer Who Wants to Change Things," *Mechanical Engineering* 88, no. 6 (June 1966): 145. Italics mine.

74. Layton, *Revolt of the Engineers*, 1, 252.

75. Noble, *America by Design*, 63.

Chapter 5

1. John Dubbury, "The Military Industrial Complex," *IEEE Transactions on Aerospace and Electronic Systems* AES-7, no. 3 (May 1971): 429–33.

2. As Ken Alder puts it, "Engineers were *designed* to serve." Alder, *Engineering the Revolution*, 86.

3. Schevitz, *Weaponsmakers*, 5–6.

4. Bob Aldridge, "The Forging of an Engineer's Conscience," *Spark* 3, no. 2 (fall 1973): 2–5.

5. Ibid., 2.

6. Everett Mendelsohn, "The Politics of Pessimism: Science and Technology Circa 1968," in *Technology, Pessimism, and Postmodernism*, Yaron Ezrahi, Everett Mendelsohn, and Howard P. Segal, eds. (Amherst: University of Massachusetts Press, 1995), 175–216; James Ledbetter, *Unwarranted Influence: Dwight D. Eisenhower and the Military-Industrial Complex* (New Haven: Yale University Press, 2011), 106–63.

7. Seymour Melman, *Pentagon Capitalism: The Political Economy of War* (New York: McGraw-Hill, 1970), 3.

8. Mumford, *Pentagon of Power*, 256–57.

9. Marc Pilisuk and Thomas Hayden, "Is There a Military-Industrial Complex?" in Pursell, *Military-Industrial Complex*, 78–79.

10. Marcuse, *One Dimensional Man*, 7.

11. Francis J. Lavoie, "The Activist Engineer: Look Who's Getting Involved!" *Machine Design* (November 2, 1972): 82–88.

12. Thorstein Veblen, *The Engineers and the Price System* (New York: Huebsch, 1921), 135.

13. Perrucci and Gerstl, *Profession without Community*, 27–89; Everett Carll Ladd Jr. and Seymour Martin Lipset, "Politics of Academic Natural Scientists and Engineers," *Science* 176, no. 4039 (June 9, 1972): 1091–1100.

14. Among members of SftP, for example, engineers oscillated between allies and scapegoats. Bill Zimmerman et al., *Toward a Science for the People* (Boston: New England Free Press, 1972). A far more tempered but nonetheless similar conflict over the relationship between scientists and technology took place among members of the president's Science Advisory Committee. Zuoyue Wang, *In Sputnik's Shadow: The President's Science Advisory Committee and Cold War America* (New Brunswick, NJ: Rutgers University Press, 2008), 258–310.

15. Lavoie, "Activist Engineer," 88.

16. Gary Benensen, "Engineering Unionism: A Recent History," *Spark* 3, no. 2 (fall 1973): 6–9; David Westman, "News from SESPA," *Spark* 2, no. 2 (fall 1972): 22.

17. Jonathan Allen, ed., *March 4: Scientists, Students, and Society* (Cambridge: MIT Press, 1970); Dorothy Nelkin, *The University and Military Research: Moral*

Politics at MIT (Ithaca: Cornell University Press, 1972); Leslie, *The Cold War and American Science*, 233–56.

18. Massachusetts Institute of Technology, *Review Panel on Special Laboratories: Final Report* (Cambridge: MIT, 1969), 73. See also Nelkin, *University and Military Research*, 71.

19. Union of Concerned Scientists, "Faculty Statement," in Allen, *March 4*, xxii–xxv.

20. Victor F. Weisskopf, "Intellectuals in Government," in Allen, *March 4*, 25–29. Rebecca Slayton, "Speaking as Scientists: Computer Professionals in the Star Wars Debate," *History and Technology* 19, no. 4 (2003): 335–64.

21. Joel Feigenbaum, "Students and Society," in Allen, *March 4*, 2–7; Moore, *Disrupting Science*, 133–46.

22. William F. Buckley Jr. "What's behind 'March 4—MIT,'" *Los Angeles Times* (February 17, 1969): B8.

23. "Research: A Policy of Protest," *Time* (February 28, 1969): 60.

24. McMahon, *Making of a Profession*, 253–56; Union of Concerned Scientists, "Forum: Concerned Scientists," *IEEE Spectrum* 6, no. 4 (April 1969): 8; M.R. Heembrock, "Forum: Concerned Scientists," *IEEE Spectrum* 6, no. 7 (July 1969): 8; Seymour Tilson, "Report on the ABM," *IEEE Spectrum* 6, no. 8 (August 1969): 24–39; George Rathjens, "Against the ABM," *IEEE Spectrum* 6, no. 8 (August 1969): 40–45; Donald Brennan, "For the ABM," *IEEE Spectrum* 6, no. 8 (August 1969): 46–50; F. K. Willenbrock, "Forum: Open Letter to Members of IEEE," *IEEE Spectrum* 6, no. 9 (September 1969): 6; J. J. G. McCue, "Spectral Lines," *IEEE Spectrum* 6, no. 10 (October 1969): 27.

25. David Dayton, "Problems and Possibilities in Reconversion," in Allen, *March 4*, 48.

26. Ronald Probstein, "Reconversion and Academic Research," in Allen, *March 4*, 36.

27. "The Reconversion to an Environmental Program of Research, Education and Public Service," December 1970, and J. A. Fay, J. C. Keck, and R. F. Probstein to H. W. Johnson and J. B. Wiesner, memorandum, 8 December 1971, box 9, folder "Fluid Mechanics Laboratory," Department of Mechanical Engineering, Records, 1942–1992 (AC259), MIT Institute Archives and Special Collections [hereafter cited as MIT ME Records].

28. See, for example, James A. Fay and D. Golomb, *Energy and the Environment* (New York: Oxford University Press, 2002).

29. Wallace D. Hayes and Ronald F. Probstein, *Hypersonic Flow Theory* (New York: Academic Press, 1959), vii–viii, 1.

30. Ibid., 6.

31. James A. Fay, *Physical Processes in the Spread of Oil on a Water Surface*, (Springfield, VA: National Technical Information Service, 1971).

32. Probstein, "Reconversion and Academic Research," 34–35.

33. Ronald F. Probstein, "Reconversion and Non-military Research Opportunities," *Astronautics and Aeronautics* 7, no. 10 (October 1969): 50–56.

34. Ibid., 50.

35. Probstein, "Reconversion and Academic Research," 34.

36. Probstein, "Reconversion and Non-military Research Opportunities," 51.

37. Ronald F. Probstein, "Taking Effective Federal Action on the Environment," *Astronautics and Aeronautics* 9, no. 1 (January 1971): 15, 80; *Application of Aerospace and Defense Industry Technology to Environmental Problems: Hearings before the Subcommittee on Government Operations House of Representatives*, 91st Cong., 74–77 (1970) (Statement of Dr. Probstein).

38. Ascher H. Shapiro, "A Position Paper on Retention or Divestiture of the Special Laboratories," 2 February 1970, box 2, folder "Scrapbook," Ascher Shapiro Papers (MC387), MIT Institute Archives and Special Collections. See also Leslie, *Cold War and American Science*, 238–39.

39. *Application of Aerospace and Defense Industry Technology to Environmental Problems: Hearings before the Subcommittee on Government Operations House of Representatives*, 91st Cong., 74–77 (1970) (Statement of Dr. Probstein).

40. "Discussion," in Allen, *March 4*, 54–56.

41. Robert L. Herbst, "War Mobilization Rally to Promote NY March," *Daily Princetonian* (April 12, 1967): 1, 3.

42. Amy Sue Bix, "'Backing into Sponsored Research': Physics and Engineering at Princeton University, 1945–1970," *History of Higher Education Annual* (1993): 9–52.

43. Robert Kargon and Stuart W. Leslie, "Imagined Geographies: Princeton, Stanford and the Boundaries of Useful Knowledge in Postwar America," *Minerva* 32, no. 2 (June 1994): 121–43.

44. Michael S. Mahoney, "'Engineering *Plus*': Technical Education and the Liberal Arts at Princeton," in *From College to University: Essays on the History of Princeton University*, ed. Anthony T. Grafton and John M. Murrin (web version 1996) Accessed October 15, 2011. http://www.princeton.edu/~hos/Mahoney/seashist/engineer.htm.

45. Ibid. Says Mahoney, "it was the U.S. government, rather than industry, that provided the bulk of the funding that increasingly characterized a new way of life for the School of Engineering. . . . By 1967, the research budget reached $3.5 million, $3.1 million of it from outside sponsors. The Dean estimated that 'more than three-quarters of the Engineering Faculty are actively engaged in research and that the programs of at least half the faculty, and more than half of the Engineering graduate students are supported by sponsored research contracts and grants.'"

46. Alexander J. Smits and Courtland D. Perkins, "Aerospace Education and Research at Princeton University, 1942–1975" Accessed October 1, 2011. http://www.princeton.edu/mae/aboutus/history-1/.

47. Mahoney, "Engineering Plus."

48. Steve M. Slaby, *The Labor Court in Norway* (Oslo: Norwegian Academic Press, 1952).

49. Minutes of the Committee on Freshman Year, 3 June 1957, box 4, folder 3, Steve M. Slaby Papers, Seeley G. Mudd Manuscript Library, Princeton University [hereafter Slaby Papers].

50. Steve Slaby, "Statement to the Engineering Faculty," 17 May 1960, box 4, folder 3, Slaby Papers.

51. Interoffice Correspondence to Prof. Glassman, 6 February 1964, box 1, folder 1, Slaby Papers.

52. Steve Slaby to Dean Elgin, 8 November 1965, box 1, folder 5, Slaby Papers.

53. Steve Slaby to the Committee on Freshman Year, memo, 20 December 1961, box 4, folder 4, Slaby Papers.

54. "proposal for a new freshman course in engineering," box 2, folder 1, and Steve Slaby to Dean Elgin, box 4, folder 3, Slaby Papers.

55. Steve Slaby to Dean Brown, 1961, box 1, folder 17, "Image of Engineering," 13 May 1963, box 1, folder 5, Slaby Papers.

56. Steve Slaby to Renato Pasta, 6 February 1970, box 1, folder 15, Slaby Papers; Steve M. Slaby, "Technology and Society: Present and Future Challenges," in *Man-Society-Technology: Representative Addresses and Proceedings of the American Industrial Arts Association 32 Annual Convention*, Linda A. Taxis, ed. (Washington, DC: American Industrial Arts Association, 1970), 9–11.

57. Rebecca Press Schwartz, "Re-evaluating Sponsored Research at Princeton: the Kuhn Committee and IDA" (April 10, 2000), unpublished manuscript.

58. John Merritt McEnany, *The Princeton Strike of May 1970: A Narrative and Commentary* (undergraduate thesis, Princeton, 1972), 24.

59. Steve Slaby to Edward Farkas, 14 Feburary 1967, box 1, folder 1, Slaby Papers.

60. Steve Slaby to Lyman Spitzer 5 November 1969, box 1, folder 1, Slaby Papers.

61. McEnany, *Princeton Strike*, 41.

62. Ibid., 48.

63. Jerome B. Simandle, "Today's Engineering Students," *Professional Engineer* 41 (May 1971): 39–41.

64. Steve M. Slaby, "What Should We Ask of the History of Technology?" in *The History and Philosophy of Technology*, George Bugliarello and Dean B. Doner, eds. (Urbana: University of Illinois, 1979), 126.

65. "Engineering Crisis" flyer, 1971, box 4, Paschkis Papers.

66. Committee for Social Responsibility in Engineering, "Statement of Purpose," *Spark* 1, no. 1 (spring 1971): 3.

67. Karl D. Stephan, "Notes for a History of the IEEE Society on Social Implications of Technology," *IEEE Technology and Society Magazine*, 25, no. 4 (2006): 5–14.

68. In 1957, Lyttle gave up his medical research equipment business when jailed for refusing to report for military service. Brad Lyttle, "Using Engineering in the Movement," *Spark* 4, no. 1 (spring 1974): 10–15.

69. Amitai Etzioni, "Minerva: An Electronic Town Hall," *Policy Sciences* 3, no. 4 (1972): 457–74.

70. Frank Collins, "Politics in the Technical Societies," *Spark* 2, no. 1 (spring 1972): 4–5. Paul Stoller and Ted Werntz, "Extraordinary Qualifications, But in Whose Interests?" *Spark* 3, no. 1(spring 1973): 14–15.

71. Stephen Unger, "Employee-Professional," *Spark* 2, no. 1 (spring 1972): 2–3.

72. Anthony Robbi, "The Structure of IEEE," *Spark* 1, no. 1 (spring 1971): 8; Harvey Rubin, "Additional Roles for IEEE," *Spark* 1, no. 1 (spring 1971): 17; "Petition for Amendment to the IEEE Constitution," *Spark* 1, no. 2 (fall 1971): 21.

73. John Harris, "On Doing It at the Work Place" *Spark* 1, no. 2 (fall 1971): 14–15; "Electronic Control of Deviant, Dissident and Delinquent Americans," *Spark* 1, no. 2 (fall 1971): 26.

74. Edward G. Trunk, "Pacifica Engineering" *Spark* 1, no. 1 (spring 1971): 39.

75. "Employment Clearing House," *Spark* 1, no. 1 (spring 1971): 30–31; Harris "On Doing It," 14.

76. "Supply and Demand Economics Are Great as Long as You the Supply Are in Demand," *Spark* 1, no. 1 (spring 1971): 50.

77. Lee M. Horowitz, "In Opposition to the Amendment" *Spark* 1, no. 2 (fall 1971): 21.

78. Aquarius Project, "Revolutionary Engineering," *Spark* 1, no. 2 (spring 1971): 12.

79. Council of Engineers and Scientists Organizations, "Administration Concerned about Unemployment Affecting Engineers and Scientists in the United States," reprinted in *Spark* 1, no. 1 (spring 1971): 9–11.

80. Murray Beaver, "Dear Spark," *Spark* 3, no. 1 (spring 1973): 2.

81. Denis Brasket, "Testimony of Denis Brasket," *Spark* 2, no. 2 (fall 1972): 12–15.

82. Paul Stoller and Ted Werntz, "The NARMIC Air War Slide Show," *Spark* 2, no. 2 (fall 1972): 16–17.

83. Paul Stoller and Ted Werntz, "Why Does DOD Pressure Engineering Unions to Promote War Bonds?" *Spark* 3, no. 2 (fall 1973): 12–14.

84. Jerome B. Wiesner, "Human Communications," *IEEE CSIT Newsletter* 4 (September 1973): 3–5.

85. "The Engineer and Military Technology," *IEEE CSIT Newsletter* 5 (December 1973): 14–17; "IEEE Disruption Exercise," *Spark* 3, no. 1 (spring 1973): 18–19.

86. Stephen H. Unger, "Codes of Engineering Ethics," *IEEE CSIT Newsletter* 5 (December 1973b): 1, 4–5.

87. Stephen H. Unger, "The BART Case: Ethics and the Employed Engineer," *IEEE CSIT Newsletter* 4 (September 1973a): 6–8.

88. R. J. Bogumil, "Social Implications of Technology: The Past and Future," *IEEE CSIT Newsletter* 9, no. 3 (September 1981): 1, 19–24.

89. *IEEE Technology and Society Magazine* 27, no. 4 (winter 2008).

90. Simon Ramo, "Toward a Social-Industrial Complex," *The Wall Street Journal* (February 16, 1972a): 10.

91. For one of the few full-throated defenses see John Stanley Baumgartner, *The Lonely Warriors: Case for the Military-Industrial Complex* (Los Angeles: Nash Publishing, 1970).

92. "CSRE Spring–Summer Offensive, 1971," *Spark* 1, no. 2 (fall 1971): 8–9.

93. On the performative elements of radicalism, see Moore, *Disrupting Science*, 158–89.

94. "The 1971 IEEE: Not Bad, Considering," *Electronics* 4, no. 8 (April 12, 1971): 37–38.

95. "Is Your Corporation Helping Generate Another Vietnam?" *Spark* 3 no. 2 (fall 1973): 16–17.

96. Fradin argued that: "We are the serious students, the ones who go to class when others seek to shut down the school. . . . We will not give up our dreams of peace, clean environment and social progress brought about with the aid of aerospace." "Supersonic Counterattack," *Time* (March 22, 1971): 15.

97. Federation of Americans Supporting Science and Technology (Summer, 1973) pamphlet; *FASST News* (July 1973), box 1, folder "T&S Correspondence," MIT Program in Science, Technology, and Society (AC363), MIT Institute Archives and Special Collections [hereafter cited as MIT STS Records].

98. "SESPA Tells It Like It Is: Opening Statement at AAA$ 1970," *Science for the People* 3, no. 1 (1971): 6–8.

99. Leventman, *Professionals Out of Work*, 189.

100. Stephen H. Unger, *Controlling Technology Ethics and the Responsible Engineer,* 2nd ed. (New York: Wiley, 1994).

101. Chris Murray, "From GE to the Syracuse Peace Council," *Spark* 4, no.1 (spring 1974): 2–3.

102. Ibid., 3.

Chapter 6

1. Donald A. Schön, *Technology and Change: The New Heraclitus* (New York: Delacorte Press, 1967), 202.

2. Kurt Vonnegut Jr., *Player Piano* (New York: Delacorte Press, 1952), 6, 27, 61, 77, 91. Other historians have found similar insight in Vonnegut's story. See especially Bess Williamson, "Small Scale Technology for the Developing World:

Volunteers for International Technical Assistance, 1959–1971," *Comparative Technology Transfer and Society* 6, no. 3 (December 2008): 236–58; Howard P. Segal, *Future Imperfect: The Mixed Blessings of Technology in America* (Amherst: University of Massachusetts Press, 1994), 126–46. For my initial thoughts on Vonnegut see Matthew Wisnioski, *Engineers and the Intellectual Crisis of Technology* (PhD dissertation, Princeton, 2005), 115–17.

3. H. G. Rickover, "Needed: A Humanistic Technology," *Professional Engineer* 39, no. 5 (May 1969): 35–37.

4. William Leavitt, "Toward a Humane Technology," *Air Force/Space Digest* 52, no. 5 (May 1969): 64–66, 71.

5. Florman, *Existential Pleasures*, 126.

6. Carroll Pursell, "The Rise and Fall of the Appropriate Technology Movement in the United States, 1965–1985," *Technology and Culture* 34, no. 3 (July 1993): 629–37; "Carroll Pursell, "The History of Technology as a Source of Appropriate Technology," *The Public Historian* 1, no. 2 (winter 1979): 15–22; Witold Rybczynski, *Paper Heroes: A Review of Appropriate Technology* (Garden City, NY: Anchor Books, 1980).

7. Guy Ortolano, *The Two Cultures Controversy: Science, Literature and Cultural Politics in Postwar Britain* (Cambridge: Cambridge University Press, 2009); D. Graham Burnett, "A View from the Bridge: The Two Cultures Debate, Its Legacy, and the History of Science," *Daedalus* 128, no. 2 (1999): 193–218.

8. Winner, *Autonomous Technology*, 55–57.

9. R. Buckminster Fuller, "Commitment to Humanity," *The Humanist* 30 (May/June 1970): 28–33.

10. E. F. Schumacher, *Small Is Beautiful: Economics as If People Mattered* (New York: Harper and Row, 1973), 138–42.

11. Turner, *From Counterculture to Cyberculture*, 2, 16, 25–26, 104; see also David Kaiser, *How the Hippies Saved Physics: Science, Counterculture, and the Quantum Revival* (New York: Norton, 2011).

12. In one extreme case the Ampex engineer Myron Stolaroff became an evangelist for LSD, turning-on Bay Area technical workers to aid in design visualization by overcoming "Midwest engineer's syndrome." John Markoff, *What the Dormouse Said: How the Sixties Counterculture Shaped the Personal Computer Industry* (New York: Viking, 2005), 65–66.

13. RWF, "Recruitment and Human Values," *Chemical Engineering Education* 3, no. 1 (winter 1969): 3.

14. Jamie Cohen-Cole, "The Creative American: Cold War Salons, Social Science, and the Cure for Modern Society," *ISIS* 100, no. 2 (June 2009): 219–62.

15. J. H. McPherson, "How to Manage Creative Engineers," *Mechanical Engineering* 87, no. 2 (February 1965): 32–36; Deutsch and Shea, Inc., *Company Climate and Creativity: 105 Outstanding Authorities Present Their Views* (New York: Industrial Relations News, 1959); Donald S. Pearson, *Creativeness for Engineers: A Philosophy and a Practice* (University Park: Pennsylvania State

University, 1958); Eugene K. Von Fange, *Professional Creativity* (Englewood Cliffs, NJ: Prentice-Hall, 1959).

16. Harold R. Buhl, *Creative Engineering Design* (Ames, IA: Iowa State University Press, 1960), 35.

17. Peter Galison, "The Americanization of Unity," *Daedalus* 127, no. 1 (winter 1998): 45–71.

18. Melvin Kranzberg, "Technology and Human Values," *Virginia Quarterly Review* 40 (Autumn 1964): 578–92; Emmanuel G. Mesthene, "Technology and Human Values," *Science Journal* 5A, no. 4 (October 1969b): 45–50.

19. General Electric Company, "Dan Johnson Has a Flair for Making Things," *Spartan Engineer* 21 no. 4 (May 1968): 26.

20. Benjamin W. Roberts and Walter Lowen, "VITA: Solving Technical Design Problems throughout the World," *Engineering Education* 58, no. 7 (March 1968): 815–17.

21. Edward S. Dennison, "Technology Transfer within Developing Areas," in *Metrology and Standardization in Less Developed Countries: The Role of a National Capability for Industrializing Economies*, H. L. Mason and H. S. Peiser eds. (Washington, DC: National Bureau of Standards, 1971), 297.

22. Williamson, "Small Scale Technology," 239.

23. Robert M. Goldhoff, "VITA and Materials Technology: Coupling Technical Skills with Problems of World Development," *An Interamerican Approach for the Seventies*, vol. 2, *Materials Education*, Southwest Research Institute ed. (New York: ASME, 1970), 29.

24. Williamson, "Small Scale Technology," 236–37.

25. Joshua Lerner, *Volunteers in Technical Assistance: The First Twenty-Five Years of VITA* (St. Louis: Center for Development Technology, Washington University, 1984), 1–5.

26. Harold L. Hoffman, "VITA—A Sociotechnical Challenge for the IEEE," *IEEE Spectrum* 5, no. 8 (August 1968): 82–89.

27. Ethan Barnaby Kapstein, "The Solar Cooker," *Technology and Culture* 22, no. 1 (January 1981): 112–21.

28. Lerner, *Volunteers in Technical Assistance*, 11–19.

29. Chauncey Guy Suits quoted in Lerner, *Volunteers in Technical Assistance*, 16.

30. J. Herbert Hollomon, "Engineering's Great Challenge—The 1960s," in *Engineering as a Social Enterprise*, Hedy E. Sladovich, ed. (Washington, DC: National Academy Press, 1991), 104–10; Lerner, *Volunteers in Technical Assistance*, 16, 24.

31. Ibid., 25–26.

32. James Daniel, "VITA Has the Answer," *Reader's Digest* 87, no. 522 (October 1965): 123–26.

33. Hoffman, "Sociotechnical Challenge," 82–89.

34. Dora Merris, "Foreign Aid Experiment Attracts Volunteer Engineers," *Product Engineering* (February 19, 1962): 114–16. Lerner, *Volunteers in Technical Assistance*, 20.

35. See, for example, "Individual's Knowledge is Utilized on an International Scale by VITA," *Chemical Engineering* 70, no. 19 (September 16, 1963): 76; "'VITA'—The Volunteers for International Technical Assistance, Inc." *RCA Engineer* 10, no. 4 (December 1964/January 1965): 88; "VITA: Expert at Exporting Ideas," *Instrumentation Technology* 17, no. 1 (January 1970): 10, 12.

36. Roberts and Lowen, "VITA," 815.

37. Lerner, *Volunteers in Technical Assistance*, 21–23.

38. Robert P. Morgan, "Applying Technology to a Social Context: Education, Housing," *Technos* 1 (January/March 1972): 49–54.

39. Roberts and Lowen, "VITA," 816.

40. "Creative Engineering," *Industrial Water Engineering* 5, no. 8 (August 1968a): 26–27.

41. Maynard Charles Nugent, "Community Service and Urban Problems: Can Civil Engineers Make a Contribution?" *Civil Engineering* 40, no. 1 (January 1970): 53; Hoffman, "Sociotechnical Challenge," 82.

42. Donald Fitzgerald, "Volunteers for International Technical Assistance," *Spark* 2, no. 1 (spring 1972): 21.

43. Brendan Jones, "Developing Lands Call on 'Brain Bank,'" *New York Times* (July 6, 1969): F12.

44. "Village Technology," *Whole Earth Catalog* (Fall 1968): 18; Dennison, "Technology Transfer," 297.

45. Dennison, "Technology Transfer," 300.

46. Goldhoff, "VITA and Materials Technology," 27; James D. Palmer, "Informed or Novice: The Engineer in Technology Transfer," in *1971 IEEE International Convention Digest* (New York: IEEE, 1971): 266–67.

47. Merris, "Foreign Aid Experiment," 115.

48. Dennison, "Technology Transfer," 300; Goldhoff, "VITA and Materials Technology," 31–32.

49. Hoffman, "Sociotechnical Challenge," 85.

50. William N. Ellis, "AT: The Quiet Revolution," *Bulletin of the Atomic Scientists* 33, no. 9 (November 1977): 24–29.

51. Lerner, 46–50.

52. "VITA Broadens Its Scope to Tackle Domestic Problems" *Product Engineering* 41, no. 14 (July 6, 1970): 21–24.

53. Lerner, *Volunteers in Technical Assistance*, 47–49.

54. "VITA Broadens its Scope," 21; Volunteers in Technical Assistance, *A Guide to Mobilizing Technical Assistance Volunteers: A 'How-to' Manual for Organizing Local Volunteer Assistance for Community Problem Solving*" (Mt. Rainier, MD: VITA, 1975), 1.

55. Lerner, *Volunteers in Technical Assistance*, 51; The Great Atlantic and Pacific School Conspiracy, *Doing Your Own School: A Practical Guide to Starting and Operating a Community School* (Boston: Beacon Press, 1972), 138; Jonathan Kozol, *Free Schools* (Boston: Houghton Mifflin, 1972), 140.

56. Lerner, *Volunteers in Technical Assistance*, 52–57.

57. Lerner, *Volunteers in Technical Assistance*, 59–60.

58. Richard Morse, *Responding to Technical Information Needs in Developing Countries: An Evaluative Review of the VITA International Inquiry Service* (Washington, DC: AID, 1972), 116–23.

59. Lerner, *Volunteers in Technical Assistance*, 74–77.

60. *Appropriate Technology: Hearings before the Subcommittee on Domestic and International Scientific Planning, Analysis, and Cooperation of the Committee on Science and Technology, US House of Representatives.* 95th Cong., 122–24 (1978) (Statement of Thomas H. Fox).

61. Jones, "Developing Lands," F12.

62. *Appropriate Technology: Hearings*, 124, 132–37.

63. Lerner, *Volunteers in Technical Assistance*, 84–101.

64. Nilo Lindgren, "Art and Technology: II. A Call for Collaboration," *IEEE Spectrum* 6, no. 5 (May 1969b): 46–56. Nilo Lindgren, "Art and Technology: I. Steps toward a New Synergism," *IEEE Spectrum* 6, no. 4 (April 1969a): 59–68.

65. Jon E. Browning, "Engineer-Artist Teams Shape New Art Forms," *Chemical Engineering* (February 26, 1968): 102–104.

66. Lindgren, "Art and Technology: II.," 56.

67. Jill Johnson, "Post Mordem," *Village Voice* (December 15, 1966): back cover.

68. Anne Collins Goodyear, "Gyorgy Kepes, Billy Klüver, and American Art of the 1960s: Defining Attitudes toward Science and Technology" *Science in Context* 17, no. 4 (2004): 611–35; Caroline A. Jones, *Machine in the Studio: Constructing the Postwar American Artist* (Chicago: University of Chicago Press, 1996); Douglas Davis, *Art and the Future: A History-Prophecy of the Collaboration between Science, Technology and Art* (New York: Praeger, 1973).

69. Alvin S. Weinstein and Stanley W. Angrist, *An Introduction to the Art of Engineering* (Boston: Allyn and Bacon, 1970), 1–7.

70. Cutler Hammer, "Art at AIL," *IEEE Spectrum* 4, no. 5 (1967): 5.

71. At the Royal Institute he completed his senior thesis, a film about the motion of electrons in electric and magnetic fields, under Nobel prize winning physicist Hannes Alfvén. Paul Miller, "The Engineer as Catalyst: Billy Klüver on Working with Artists," *IEEE Spectrum* 35, no. 7 (July 1998): 20–29.

72. Billy Klüver, "The Garden Party," in *The Machine: As Seen at the End of the Mechanical Age*, K. G. Pontus Hultén, ed. (New York: Museum of Modern Art, 1968), 169–71.

73. Billy Klüver, "Fragment on Man and the System," *The Hasty Papers: A One Shot Review* (New York: Alfred Leslie, 1960): 45.

74. Fred Waldhauer as quoted in Brian O'Doherty, "At the '10th evening meeting,'" 30 November 1966, box 1, folder 43, Experiments in Art and Technology Records, 1966–1993, 940003. Getty Research Institute [hereafter cited as E.A.T. Records].

75. Billy Klüver and Robert Rauschenberg, "The Organization of E.A.T.," box 3, folder 17, E.A.T. Records.

76. Julie Martin, "Program" *E.A.T. News*, 1, no. 3 (November 1967a): 6.

77. Billy Klüver, "Dear Elenore," 28 August 1966, box 3, folder 1, E.A.T. Records.

78. Billy Klüver, talk before the IEEE New Technical and Scientific Activities Committee, 21 March 1967, box 125, folder 49, E.A.T. Records; Charles Cook to Billy Klüver, 24 March 1967, box 125, folder 49, E.A.T. Records; Billy Klüver to J. B. Fisk, 1967, box 3, folder 2, E.A.T. Records; J. R. Pierce to Billy Klüver, 1967, box 3, folder 2, E.A.T. Records.

79. Fred Turner, "Romantic Automatism: Art, Technology, and Collaborative Labor in Cold War America," *Journal of Visual Culture* 7, no. 1 (2008), 21–23.

80. "Automation House: a Philosophy for Living in a World of Change," *New York Times* (February 1, 1970): AS 2. The draft copy of the statement had two paragraphs summarizing Mesthene's work at the Harvard Program. "New York Times Supplement," box 139, folder 12, E.A.T. Records.

81. Billy Klüver to Julius Stratton, 10 June 1968, box 42, E.A.T. Records; Billy Klüver, "Plans for Mobilizing the Technical Community," box 120, folder 6, E.A.T. Records.

82. Billy Klüver to Emmanuel Mesthene, 10 June 1968, box 122, folder 2, E.A.T. Records; Billy Klüver, "The Ghetto and the Technical Community: An Opportunity for Collaboration," box 120, folder 9, E.A.T. Records.

83. memo, 17 March 1967, box 3, folder 17, E.A.T. Records; "policy for E.A.T. national organization," box 29, folder 1, E.A.T. Records.

84. Julie Martin, *E.A.T. News* 1, no. 4 (December 20, 1967b): 3.

85. "E.A.T. Has a Project for You," *Ampex This Week*, 11 June 1969, box 29, folder 13, E.A.T. Records.

86. Engineer information cards, box 8, folders 1–10, E.A.T. Records; H. B. Newman, engineer information card, box 8, folder 5, E.A.T. Records.

87. Bruce Steinberg to Billy Klüver and Robert Rauschenberg, 18 March 1968, box 10, folder 4, E.A.T. Records.

88. Questionnaire to participants of *9 Evenings*, box 1, folder 38, E.A.T. Records; L. J. Robinson, "On Art and Technology," May 1967, box 121, folder 11, E.A.T. Records.

89. Donna Wilson to Joyce Libby, 15 July 1969, box 6, folder 26, E.A.T. Records.

90. John Pruner to E.A.T., box 6, folder 26, E.A.T. Records.

91. J. D. McGee and L. J. Heilos, "Visual Display of Infrared Laser Output on Thermographic Phosphor Screens," *IEEE Journal of Quantum Electronics* 3, no. 1 (1967): 31.

92. Manfred R. Schroeder, "Images from Computers," *IEEE Spectrum* 6, no. 3 (March 1969): 66–78.

93. J. R. Pierce to W. O. Baker, 4 April 1966, box 4, folder 2, E.A.T. Records.

94. Nilo Lindgren, "Into the Collaboration," in *Pavilion*, Billy Klüver, Julie Martin, and Barbara Rose, eds. (New York: Dutton, 1972), 3–59.

95. Billy Klüver, "The Pavilion," in Klüver et al., *Pavilion*, ix.

96. Fujiko Nakaya, "Making of 'Fog' or Low-Hanging Stratus Cloud," in Klüver et al., *Pavilion*, 207–23; Thomas R. Mee, "Notes and Comments on Clouds and Fog," in Klüver et al., *Pavilion*, 224–27.

97. E.A.T. claimed the mirror had a more precise optical surface than a communications satellite. Lindgren, "Into the Collaboration," 39.

98. Alan Pottasch to Billy Klüver, 19 March 1970, box 43, folder 31, E.A.T. Records.

99. Billy Klüver to Donald Kendall, 8 April 1970, box 122, folder 14, E.A.T. Records.

100. Gordon Friedlander, "Art and Technology: A Merger of Disciplines," *IEEE Spectrum* 6, no. 10 (December 1969): 60–68.

101. Alex Gross, "Who Is Being Eaten?" *East Village Other* (March 3, 1970): 14; Anne Collins Goodyear, "From Technophilia to Technophobia: The Impact of the Vietnam War on the Reception of 'Art and Technology,'" *Leonardo* 41, no. 2 (April 2008): 169–73.

102. Executive meeting minutes, 12 September 1970, box 128, folder 29, E.A.T. Records; Billy Klüver, notes, box 53, folder 13, E.A.T. Records.

103. Paul Goodman, "Can Technology Be Humane?" *New York Review of Books* (November 20, 1969b): 27–34.

104. Ruth Schwartz Cowan, "Looking Back in Order to Move Forward: John McDermott, 'Technology: The Opiate of the Intellectuals,'" *Technology and Culture* 51, no. 1 (January 2010): 199–215.

105. Paul Goodman, "The Case Against Technology," *Innovation* 2 (June 1969a): 36–47.

106. "The Innovation Group!" *Scientific American* 220, no. 6 (June 1969): 125; "The Innovation Group!" *Wall Street Journal* (April 25, 1969): 4. *Innovation* was roughly four times as expensive as an engineering society membership and five times as much as the famously pricey *Fortune* magazine.

107. Patrick McCurdy, "The Innovative Group Innovates with TV," *Chemical Engineering News* 48, no. 7 (February 16, 1970): 16–17; Michael F. Wolff, "Says the Editor," *Innovation* 27 (January 1972): 1.

108. "Retrieval," *Innovation* 23 (August 1971): 65.

109. Dorothy Nelkin, *The Politics of Housing Innovation: The Fate of the Civilian Industrial Technology Program* (Ithaca: Cornell University Press, 1971), 18–19.

110. Ibid., 32.

111. Hollomon, "Engineering's Great Challenge," 105.

112. Engineers Joint Council, *The Nation's Engineering Research Needs 1965–1985* (New York: EJC, 1962).

113. Arthur D. Little, Inc., *Patterns and Problems of Technical Innovation in American Industry* (Washington, DC: US Government Printing Office, 1963).

114. Nelkin, *Politics of Housing Innovation*, 28–46.

115. Daniel V. DeSimone, introduction to *Education for Innovation*, Daniel V. DeSimone, ed. (Oxford: Pergamon Press, 1968), vii; J. Herbert Hollomon, "Creative Engineering and the Needs of Society," in DeSimone, *Education for Innovation*, 23–30.

116. See, for example, The National Industrial Conference Board, ed., *The Challenge of Technology: Linking Business, Science, and the Humanities in Examining Management and Man in the Computer Age* (New York: The Conference Board, 1966).

117. James D. O'Connell, Eugene G. Fubini, Kenneth G. McKay, James Hillier, J. Herbert Hollomon, "Electronically expanding the citizen's world," *IEEE Spectrum* 6, no. 7 (July 1969): 30–40.

118. William G. Maass, "New Information Services from a Not-So-Old Publishing House," *Journal of Chemical Documentation* 2, no. 1 (1962): 46–48; William G. Maass, "From the Publisher: A Word of Introduction," *International Science and Technology* 1 (January 1962): front insert.

119. William G. Maass, "From the Publisher: After a Year" *International Science and Technology* 12 (December 1962): front insert.

120. David Allison, "The University and Regional Prosperity," *International Science and Technology* 40 (April 1965): 22–31; David Allison, "The Science Entrepreneur," *International Science and Technology* 13 (January 1963): 40–45.

121. Jack A. Morton, "From Research to Technology," *International Science and Technology* 29 (May 1964): 82–92.

122. Jack A. Morton, "The Microelectronics Dilemma," *International Science and Technology* 55 (July 1966): 35–44; Allison prefaced Morton's article with the warning that "This is a story of change—how to survive it and how to grow with it. And perhaps this is also a story of how a society can be great." David Allison, "In Our Opinion," *International Science and Technology* 55 (July 1966): 23; it is possible that this work was Morton's effort to understand how the transistor market had gotten away from Bell Labs. Ross Knox Bassett, *To the Digital Age: Research Labs, Start-Up Companies, and the Rise of MOS Technology* (Baltimore: Johns Hopkins University Press, 2002), 55–56.

123. The full advisory board consisted of *Robert M. Adams*, venture capitalist of New Ventures Division at WR Grace & Co., former vice president at 3M; *Warren G. Bennis*, provost at the State University of New York at Buffalo and former MIT Sloan School professor; *Emilio Daddario*; *Eugene G. Fubini*, private engineering consultant, former IBM vice president, and former assistant secretary of defense to JFK; *C. Lester Hogan*, president of Fairchild Camera and Instrument Camera and physicist who had been a Harvard professor and worked at Bell Labs and Motorola; *J. Herbert Hollomon*; *Koji Kobayashi*, founder and president of Japan's NEC Corporation; *Warren Kraemer*, the corporate vice president at McDonnell Douglas; *Donald G. Marquis*, psychologist and professor of industrial management at MIT who was director of the Office of Psychological Personnel in World War II; *Emmanuel G. Mesthene*; *Jack Morton*; *Gert W. Rathenau*, solid-state physicist at Philips Eindhoven; *Robert H. Ryan*, the Gulf Oil Realty executive responsible for the planning of Reston, Virginia; and *E. C. Williams*, chief scientist of the British Ministry of Power.

124. Warren Bennis, "How to Survive in a Revolution," *Innovation* 11 (March 1970): 2–9.

125. Emmanuel Mesthene and Herbert Hollomon, "The New Meaning of Social Responsibility," *Innovation* 28 (February 1972): 2–9.

126. "Announcing Innovation TeleSessions," *Innovation* 30 (April 1972): 55.

127. Bennis, "How to Survive," 8.

128. McCurdy, "Innovative Group," 16–17; A. J. Parisi, "New Kind of Conference Focuses on New Ideas in Technology," *Product Engineering* 41, no. 5 (March 2, 1970): 22–24; "Top Idea Men Trade Ideas," *Business Week* (January 31, 1970): 32–33. For a second Innovation Group workshop see Nancy Foy, "The Outer View: We're OK, Quality-of-Life-Wise," *The Computer Bulletin* 16, no. 2 (1972): 71, 74.

129. John M. Steele Jr. and Ronald Neswald, "How Science and Technology Just Might Survive in a Post-technological Age," *Innovation* 27 (January 1972): 48–57.

130. "Response," *Innovation* 3 (July 1969): 71.

131. Evan Herbert, "The Sophisticated Spokesman" *Innovation* 9 (1970): 68–69.

132. Schön, *Technology and Change*, 189–93; Donald A. Schön, *Beyond the Stable State* (New York: Random House, 1971), 10–18.

133. Donald A. Schön, "The Diffusion of Innovation," *Innovation* 6 (October 1969): 42–53.

134. Schön, *Technology and Change*, 205–18.

135. Michael Riordan, "How Bell Labs Missed the Microchip," *IEEE Spectrum* 43, no. 12 (December 2006): 36–41.

136. J. Herbert Hollomon, "Technology Policy: The US Picks Its Way," *IEEE Spectrum* 9, no. 8 (August 1972): 72–81.

137. Don Fabun, *The Children of Change* (Beverly Hills: Glencoe Press, 1969), 5–12. See also Heywood Gould, *Corporation Freak* (New York: Tower Publications, 1971).

138. George C. Beakley and Ernest G. Chilton with contributions by Michael J. Nielsen, *Design: Serving the Needs of Man* (New York: Macmillan, 1974).

139. Samuel C. Florman, *Adventure in the Arts: An Engineer's Guide to Great Books* (New York: McGraw-Hill, 1968a), 4–5.

140. Samuel C. Florman, *Engineering and the Liberal Arts: A Technologist's Guide to History, Literature, Philosophy, Art, and Music*, McGraw-Hill Series in Continuing Education for Engineers (New York: McGraw-Hill, 1968b), 2, 8.

141. Samuel C. Florman, "Creativity and the Anti-Technologists," in *Civil Engineers in the World around Us*, M. D. Morris, ed. (New York: ASCE, 1974), 18–21.

142. Samuel C. Florman, *Blaming Technology: The Irrational Search for Scapegoats* (New York: St. Martin's Press, 1981a), 80–96.

143. Samuel C. Florman, "Do It Yourself Is the Message," *New York Times Book Review* (January 25, 1981b): 3, 24.

144. Florman, *Existential Pleasures*, 91–98. In 1982 Henry Petroski took Florman's baton with a book whose very title named the stakes. Henry Petroski, *To Engineer Is Human: The Role of Failure in Successful Design* (New York: St. Martin's Press, 1985).

Chapter 7

1. George C. Beakley and H. W. Leach, *Engineering: An Introduction to a Creative Profession,* 2nd ed. (New York: Macmillan, 1972), 424.

2. MIT Department of Humanities, "Endicott Conference III: Carnegie Collaboration Seminar," 16 June 1968, box 4, folder "Carnegie Conference 1968 report," Department of Humanities, Records (AC404), MIT Institute Archives and Special Collections [hereafter cited as MIT Humanities Records].

3. Robert Hutchins, "Stamp out Engineering Schools," *Engineer* 9, no. 2 (March/April 1968): 17–19.

4. Betty M. Vetter, "The Military Draft: Educational and Occupational Deferments for Engineers," *Civil Engineering* 39, no. 5 (May 1969): 67–70.

5. Board of Directors, ASEE, "ASEE Resolution on Campus Disruption," *Engineering Education* 61, no. 1 (September/October 1970): 11.

6. Center for Policy Alternatives, *Future Directions for Engineering Education: System Response to a Changing World* (Washington, DC: ASEE, 1975), 85–86.

7. Carroll Seron and Susan Silbey, "The Dialectic between Expert Knowledge and Professional Discretion: Accreditation, Social Control, and the Limits of

Instrumental Logic," *Engineering Studies* 1, no. 2 (July 2009): 101–28. For a parallel case see: Fabio Rojas, *From Black Power to Black Studies: How a Radical Social Movement Became an Academic Discipline* (Baltimore: Johns Hopkins University Press, 2007).

8. Clark Kerr, *The Uses of the University* (Cambridge: Harvard University Press, 1963).

9. Seely, "Research, Engineering, and Science," 344–86.

10. Committee on Evaluation of Engineering Education, "Summary of the Report on Evaluation of Engineering Education," *Journal of Engineering Education* 46, no. 1 (September 1955): 25–60.

11. Benson R. Snyder, *The Hidden Curriculum* (New York: Knopf, 1970).

12. In an earlier gathering, the Endicott group referred to the ASEE as the "ASPEE—American Society for the Prevention of Engineering Education." A Report on the Endicott House Conference, 3 September 1966, box 4, folder "Carnegie Conference, 1966," MIT Humanities Records.

13. Humanistic–Social Research Project, *General Education in Engineering* (Urbana, IL: ASEE, 1956), 3–6, 94–95.

14. Bruce E. Seely, "The Other Re-engineering of Engineering Education, 1900–1965," *Journal of Engineering Education* 88, no. 3 (July 1999): 285–94.

15. American Society for Engineering Education, *Goals of Engineering Education: Preliminary Report* (Urbana, IL: ASEE, 1965), 11, 20, 21. Reception of the Report was contentious. See for example: "Goals of Engineering Education, 2: Dissenting Opinions," *Mechanical Engineering* 88, no. 1 (January 1966): 40–43.

16. ASEE Humanistic–Social Research Project, *Liberal Learning for the Engineer: Report of the ASEE Humanistic-Social Research Project* (Washington, DC: ASEE, 1968), 4–6, 32.

17. Ibid., 8–10.

18. Henry Knepler, "Engineering Education and the Humanities in America," *Leonardo* 6, no. 4 (1973): 305–309.

19. Humanistic–Social Research Project, *Liberal Learning*, 35–37.

20. Jane C. Keller and George R. Webb, "The Tulane Program on Science, Technology, and Man" *Engineering Education* 62, no. 4 (January 1972): 402–404. See also J. Gurland, "Engineering and Liberal Education," *Engineering Education* 61 no. 5 (February 1971): 426–28; Samuel T. Carpenter, "The Swarthmore Experience," *Engineering Education* 63, no. 5 (February 1973): 341–44.

21. P. B. Daitch, "Engineering and Social Values," *Engineering Education* 61, no 2 (November 1970): 125–26.

22. Gale E. Nevill Jr. and John A. O'Connor, "Unbottle Your Creative Ideas: A Cooperative Venture in Engineering and Art," *Engineering Education* 63, no. 2 (November 1972): 112–16. Other technical schools, including Caltech and the Stevens Institute of Technology, introduced artist-in-residence programs to eliminate the "tunnel vision" of modern engineers. Paul F. Miller, "Art for Engineers,"

Engineering Education 62, no. 3 (December 1971): 271–77; Lee Rosenthal, "A Course in Technology for Artists" *Leonardo*, 7 (1974): 27–29.

23. "Workshop on Social Directions of Technology," *Engineering Education* 61, no. 2 (November 1970): 129.

24. B. R. Dorsey, "The Engineer's Role in the Modern World," *Engineering Education* 60, no. 3 (November 1969): 226–28.

25. David P. Billington and Robert Mark, "Humanities in Civil Engineering," *Engineering Education* 59, no. 9 (May 1969): 1059–63.

26. Edwin J. Holstein and A. Bruce Carlson, "Engineering in its Social Context— An Experimental Graduate Course," *Engineering Education* 60, no. 3 (November 1969): 240–41.

27. James H. Black, "Engineering a Cornerstone of Our Society," *Engineering Education* 62, no. 3 (December 1971): 265–67.

28. E. V. Krick, "Engineering for Non-engineering Students," *Engineering Education* 62, no. 3 (December 1971): 252–54.

29. Seely, "SHOT," 749–60.

30. Melvin Kranzberg and Carroll W. Pursell Jr., "Technology's Challenge," in *Technology in Western Civilization*, vol. 2, Melvin Kranzberg and Carroll W. Pursell Jr., eds. (New York: Oxford University Press, 1967), 704.

31. See, for example: William Henry Davenport and Daniel Rosenthal, *Engineering: Its Role and Function in Human Society* (New York: Pergamon Press, 1967); Donald P. Lauda and Robert D. Ryan, eds. *Advancing Technology: Its Impact on Society* (Dubuque, IA: WMC Brown, 1971); Charles R. Walker, *Modern Technology and Civilization: An Introduction to Human Problems in the Machine Age* (New York: McGraw-Hill, 1962); John G. Burke, ed., *The New Technology and Human Values* (Belmont, CA: Wadsworth, 1966); Carroll W. Pursell, *Readings in Technology and American Life* (New York: Oxford, 1969); John P. Rasmussen, ed., *The New American Revolution: The Dawning of the Technetronic Era* (New York: Wiley, 1972).

32. Noel de Nevers, *Technology and Society* (Reading, MA: Addison-Wesley, 1972), 307.

33. Richard C. Dorf, *Technology and Society* (San Francisco: Boyd and Fraser, 1974), iii–24.

34. Ralph J. Smith, *Engineering as a Career* (New York: McGraw-Hill, 1969), C1–C20, 173.

35. Engineering Concepts Curriculum Project, *The Man Made World: A Course on the Theories and Techniques that Contribute to Our Technological Civilization*, vol. 1 (New York: McGraw-Hill, 1969), vi.

36. John G. Truxal, "Toward a More Humane Use of Technology," *Engineering Education* 62, no. 2 (November 1971): 93–94. On its early roots see: Engineering Concepts Curriculum Project, *Working Papers of the Summer Study Report on the Engineering Concepts Curriculum Project* (Washington, DC: Commission on Engineering Education, 1964).

37. Engineering Concepts Curriculum Project, *Man and His Technology* (New York: McGraw-Hill, 1973), xv. The *Man Made World* was abbreviated in future editions as *Man and His Technology*.

38. Ibid., 6–8.

39. Kerr, *Uses of the University*, 92.

40. Llewellyn M. K. Boelter, *Education for the Profession* (Los Angeles: Department of Engineering, UCLA, 1963).

41. Statistics gathered from University of California's *General Catalogue* from 1945 to 1955.

42. Carl W. Borgmann, *The Ford Foundation's Role in Engineering Education* (New York: Ford Foundation. 1964). UCLA's grant was one of three awarded by Ford's new engineering education program, along with MIT and the Case Institute.

43. Allen B. Rosenstein, *A Study of a Profession and Professional Education: The Final Publication and Recommendations of the UCLA Educational Development Program* (Los Angeles: School of Engineering and Applied Science University of California, 1968), II-13. The EDP had an inauspicious start. The UC System's Committee on Educational Policy challenged its necessity, suggesting that pedagogical studies detracted from scholarship, which was the mark of a "first class university." It delayed accepting the grant and suggested using the funds to recruit faculty. Frustrated by the treatment, co-PI Myron Tribus left to become Dean of Dartmouth's Thayer School. Committee on Educational Policy, "Ford Foundation Grant," 12 February 1960, and Myron Tribus to President Clark Kerr, 27 January 1960, box 8, (RS401) UCLA University Archives.

44. Allen B. Rosenstein, "Engineering Science in EE Education," in *Electrical Engineering Education: Proceedings of the International Conference at Syracuse and Sagamore*, Norman Balabanian and Wilbur R. Page, eds. (Syracuse, NY: Department of Electrical Engineering, Syracuse University, 1961), 95–98.

45. D. Rosenthal, A. B. Rosenstein, and G. Wiseman, *Information Theory and Curricular Synthesis* (Los Angeles: University of California, 1963). The process also figured prominently in Rosenstein, *Study of a Profession*, III.1–III.22.

46. D. Rosenthal, A. B. Rosenstein, and M. Tribus, "How Can the Objective of Engineering Education Be Best Achieved? *Data Link* 4, no. 4 (December 1961): 8–15.

47. Bonham Campbell, ed. *Proceedings of the Conference on the Humanities in Engineering Curriculum*, (Los Angeles: Department of Engineering, UCLA, 1963).

48. Daniel Rosenthal, "Skills in Humanities," in *Studies of Courses and Sequences in Humanities, Fine Arts and Social Sciences for Engineering Students*, EDP Humanities Subcommittee, ed. (Los Angeles: Department of Engineering, UCLA, 1963), 35–37.

49. William Henry Davenport and J. P. Frankel, *The Applied Humanities* (Los Angeles: Department of Engineering, UCLA, 1968), 87; William Henry Davenport and Daniel M. Rosenthal, *Engineering: Its Role and Function in Human*

Society (Los Angeles: Department of Engineering, UCLA, 1966), x. A year later their anthology became a textbook in Pergamon Press's new Humanities and Social Sciences series for engineers.

50. Rosenstein, *Study of a Profession*, II-11, II-15. Rosenstein found inspiration an essay by Lynn White Jr. that called for an erasure of the two cultures divide and for the aristocracy of humanism. Lynn White Jr., "Humanism and the Education of Engineers," in *Studies of Courses and Sequences in Humanities, Fine Arts and Social Sciences for Engineering Students*, EDP Humanities Subcommittee, ed. (Los Angeles: Department of Engineering, UCLA, 1963), 39–54.

51. Rosenstein, *Study of a Profession*, xi–xiv.

52. Chauncey Starr, "To the Prospective Engineering Student," in *Announcement of the College of Engineering* 7, no. 11 (1967): 5.

53. The trend toward graduate research was pronounced. In 1963 graduate students outnumbered undergraduates for the first time. By 1969 there were 168 faculty members, 135 undergraduate courses, and 163 graduate courses. Statistics gathered from University of California, *General Catalogue* from 1955 to 1969.

54. Allen Rosenstein, "Professions in Modern Society," *Engineering Education* 60, no. 10 (June 1970): 960–63.

55. National Technology Foundation Act of 1980, *Hearings on H.R. 6910, before the Subcommittee on Science, Research and Technology of the Committee on Science and Technology, 96th Cong. (1980).*

56. Center for Policy Alternatives, *Future Directions for Engineering Education*, 32.

57. Ad Hoc Committee on Potential Campus Disruptions, box L1.2, and "Drugs And The Caltech Students," box X2.6, California Institute of Technology Historical Files, California Institute of Technology Archives [hereafter cited as Caltech Historical Files]; box 1.1, Committee on the Freshman Year 1964–1967, California Institute of Technology Archives.

58. Caltech cast the humanities as a source of imagination for technical work and as valuable knowledge in its own right: "Literature is taught as literature, history as history, and philosophy as philosophy." *Bulletin of the California Institute of Technology* 46, no. 4 (1937), 8.

59. Interdisciplinary efforts would make it "especially attractive for many individuals, corporations, and foundations to make new and generous investments in the future of Caltech." Lee DuBridge, "Report of the President," *Bulletin of the California Institute of Technology* 76, no. 4 (1967): 3–4. In preparation for the campaign Caltech hired consultants to examine its reputation among corporate and civic leaders. Respondents were "vitally concerned" about social problems demanding solutions "in human terms," and viewed Caltech as weak in the humanities. Fry Consultants, "California Institute of Technology (An Image Study)," 1967, box K4.4, Caltech Historical Files.

60. Joseph Rhodes Jr. "Letters: Crisis at Caltech," *California Tech* (April 16, 1967), 2.

61. Associated Students of the California Institute of Technology, *Air Pollution Project: An Educational Experiment in Self-directed Research* (Pasadena: California Institute of Technology, 1968), 322.

62. Ad Hoc Faculty Committee on Aims and Goals, *Aims and Goals of the California Institute of Technology II. General Problems of Growth and Change of Caltech* (Pasadena: California Institute of Technology, 1969), 45.

63. Ibid., 41.

64. Ad hoc Faculty Committee on Aims and Goals, *Aims and Goals of the California Institute of Technology III. Introducing the Social and Behavioral Sciences at Caltech* (Pasadena: California Institute of Technology, 1969), 71.

65. Ad hoc Faculty Committee on Aims and Goals, *Aims and Goals of the California Institute of Technology VI. The Humanities at Caltech* (Pasadena: California Institute of Technology, 1969), 2.

66. Harvey Mudd College, *Bulletin, 1958–1959* (Claremont, CA: Harvey Mudd College, 1958).

67. In its first ten years HMC's faculty grew from 7 to 37, its student body from 48 to 283. Joseph B. Platt, *Harvey Mudd College: The First Twenty Years* (Santa Barbara, CA: Fithian Press, 1994), 194–207.

68. Joseph B. Platt and J. P. Frankel, "A Proposal to the Alfred P. Sloan Foundation," 9 October 1968, unprocessed files, George McKelvey Papers, Special Collections at the Libraries of the Claremont Colleges [hereafter cited as McKelvey Papers].

69. William H. Davenport, *The One Culture*, Unified Engineering Series (New York: Pergamon Press, 1970), xi–xiv, 87–120. While in Cambridge, Davenport met Lewis Mumford, with whom he struck a twelve-year correspondence. William H. Davenport, "William H. Davenport: Founding Faculty, Chairman of the Humanities and Social Sciences, Willard W. Keith Jr. Fellow in Humanities, 1957–1973: Oral History Interview," conducted by Enid Hart Douglass (1990), 113.

70. Joseph B. Platt and J. P. Frankel, "A Proposal to the Alfred P. Sloan Foundation," 9 October 1968, unprocessed files, McKelvey Papers.

71. Theodore Waldman, "Quest for Commonwealth, an Applied Humanities Course," in *Applying Humanities to Engineering* (Claremont, CA: Harvey Mudd College, 1971), 1–11.

72. Harvey Mudd College, *Bulletin, 1972–1973* (Claremont, CA: Harvey Mudd College, 1972), 1.

73. The MIT Course Evaluation Guide, February 1972, box 1, folder "Memoranda (Miscellaneous)," MIT Humanities Records.

74. Alfred H. Keil, "School of Engineering," *Massachusetts Institute of Technology Bulletin* 109, no. 4 (November, 1973): 117–23.

75. MIT Committee on Educational Survey, *Report of the Committee on Educational Survey to the Faculty of the Massachusetts Institute of Technology* (Cambridge: Technology Press, 1949), 4, 20–36.

76. Harold J. Hanham, "School of Humanities and Social Science," *Massachusetts Institute of Technology Bulletin* 109, no. 4 (November 1973): 171–72.

77. Roy Lamson, "MIT's Humanities Course," *Technology Review* 62, no. 7 (May 1960): 21–23, 50.

78. J. A. Stratton, "President," *Massachusetts Institute of Technology Bulletin* 99, no. 2 (November 1963), 1–47.

79. Gordon S. Brown and William W. Seifert, "School of Engineering," *Massachusetts Institute of Technology Bulletin* 98, no. 2 (November 1962): 55–76; Gordon S. Brown, "New Horizons in Engineering Education," *Daedalus* 91, no. 2 (1962): 341–61.

80. Richard M. Douglas, "Carnegie Application," 7 January 1965, box 4, folder "Carnegie Application," MIT Humanities Records.

81. Harold J. Hanham, "School of Humanities and Social Sciences," *Massachusetts Institute of Technology Bulletin* 111, no. 4 (November 1975): 241–46.

82. Steering Committee of Social Inquiry, "A Program in Social Inquiry," box 8, folder "Social Inquiry," MIT Humanities Records.

83. Commission on MIT Education, *Creative Renewal in a Time of Crisis* (Cambridge: MIT, 1971).

84. Arthur Steinberg, "A Minority View," in Commission on MIT Education, *Creative Renewal in a Time of Crisis* (Cambridge: MIT, 1971), 133–34.

85. Jerome Wiesner and Paul Gray, "President and Chancellor," *Massachusetts Institute of Technology Bulletin* 108, no. 3 (September 1973): 1–30.

86. Jerome B. Wiesner, "Letter to Howard Johnson, April 3, 1967," in *Jerry Wiesner: Scientist, Statesman, Humanist: Memories and Memoirs*, Walter A. Rosenblith, ed. (Cambridge: MIT Press, 2003), 473–75. Wiesner suggested that the college be built on the site of the former Watertown Arsenal so that it would not be subject to the existing disciplinary and research pressures of the MIT culture.

87. Alfred A. H. Keil, "Engineering at MIT: Perspectives and Prospects," 1 October 1971, box 1, folder "School of Engineering Directory," Urban Systems Laboratory, Records, 1968–1974 (AC366), MIT Institute Archives and Special Collections [hereafter cited as MIT USL Records].

88. Alfred A. H. Keil, "Developing MIT's School of Engineering to Meet a Broader Challenge," 17 January 1972, box 1, folder "School of Engineering Directory," MIT USL Records.

89. John Crocker, "Our Malaise," May 1972, box 8, folder "Our Malaise," Technology and Culture Seminar, Records, 1971–1976 (AC358), MIT Institute Archives and Special Collections [hereafter cited as T&C Seminar Records]. The Seminar was in its second incarnation. The first version of the Seminar was initiated by Crocker's predecessor Myron Bloy Jr. It ran from 1964 to 1968. See: Ilene Montana, ed., *Technology and Culture in Perspective* (Cambridge, MA: Church Society for College Work, 1967).

90. John Crocker Jr. "Report on the Luncheon," 25 January 1972, box 8, folder "1971–72 Committee," T&C Seminar Records.

91. "Technology Studies: Opportunities and Prospects," *Technology Studies Bulletin* 2 (January 1974), 1–16.

92. In its first newsletter, the group compiled and distributed a directory of over a hundred interested faculty, roughly a third of whom were engineers. "Technology Studies Directory," *Technology Studies Bulletin* 1 (May 1973): 6–37.

93. "Introduction," *Technology Studies Bulletin* 1 (May 1973): 1–5.

94. Nathan Sivin, "Introduction," *Technology Studies Bulletin* 2 (January 1974), i–iv.

95. "Professional Seminars," *Technology Studies Bulletin* 2 (January 1974), 17–20.

96. Larry Bucciarelli to Tom Sheridan, memorandum, 7 October 1974, 11 November 1974, box 2, folder "Master's Program," Technology Studies Program, Records (AC363), MIT Institute Archives and Special Collections.

97. "Progress Report," *Technology Studies Bulletin* 4 (June 1975), 4–8.

98. Instead of interdisciplinary courses, engineering majors enrolled in Economics and Psychology at a ratio of 10:1. Harold J. Hanham, "School of Humanities and Social Sciences," in *Report of the President and the Chancellor, 1977–1978: Massachusetts Institute of Technology* (Cambridge: MIT, 1978), 269–75.

99. David Dickson, "MIT Puts Some Polish on a Tarnished Image," *Nature* 281 (25 October 1979): 627–28.

100. Compiled from statistics in MIT's President's Reports from 1970 to 1980.

101. Downey and Lucena, "Knowledge and Professional Identity," 395.

102. Rustum Roy and Joshua Lerner, "The Status of STS Activities at U.S. Universities," *Bulletin of Science, Technology and Society* 3 (1983): 417–32.

103. Rosenstein, *Study of a Profession*, I-1.

104. Joseph Platt, "Profile of Harvey Mudd College, 1952–1972: Prepared for the Ford Foundation," June 1962, Unprocessed papers of George McKelvey, Special Collections at the Libraries of the Claremont Colleges.

105. Center for Policy Alternatives, *Future Directions for Engineering Education*, 50, 96.

106. Ibid, 42.

107. Stephen H. Cutcliffe, "The STS Curriculum: What Have We Learned in Twenty Years?" *Science, Technology, and Human Values* 15, no. 3 (summer 1990): 360–72.

108. For a similar trend in education aimed at racial diversity see: Slaton, *Race, Rigor, and Selectivity in US Engineering*, 113–42.

Chapter 8

1. Hughes, *Rescuing Prometheus*, 14.

2. John Dustin Kemper, *The Engineer and His Profession* (New York: Holt, Rinehart and Winston, 1975), 48–50.

3. Ibid., viii, 1–83.

4. Robert J. Baum and Albert Flores, eds. *Ethical Problems in Engineering* (Troy, NY: Center for the Study of the Human Dimensions of Science and Technology, 1978); Paul Durbin, "Engineering Ethics and Social Responsibility: Reflections on Recent Developments in the USA," *Bulletin of Science, Technology and Society* 17, no. 2/3 (1997): 77–83.

5. National Research Council Panel on Engineering Interactions with Society, *Engineering Education and Practice in the United States: Engineering in Society* (Washington, DC: National Academy Press, 1985), 45.

6. Meiksins, "Engineers in the United States," 89.

7. Gary Lee Downey, *The Machine in Me: An Anthropologist Sits among Computer Engineers* (New York: Routledge, 1998); Juan C. Lucena, *Defending the Nation: US Policymaking to Create Scientists and Engineers from Sputnik to the 'War against Terrorism'* (Lantham, MD: University Press of America, 2005).

8. Simon Ramo, *America's Technology Slip* (New York: Wiley, 1980).

9. Downey, *Machine in Me*, 6–10.

10. National Research Council Panel on Engineering Interactions with Society, *Engineering in Society*, 36–40.

11. Ibid., 16, 74.

12. Ronald R. Kline, "Cybernetics, Management Science, and Technology Policy: The Emergence of 'Information Technology' as a Keyword, 1948–1985," *Technology and Culture* 47, no. 3 (July 2006): 513–35.

13. Thomas Friedman, "It's a Flat World After All," *New York Times* (April 3, 2005): 33–37.

14. Hughes, *Rescuing Prometheus*, 301–305.

15. ASCE Steering Committee to Plan a Summit on the Future of the Civil Engineering Profession in 2025, *The Vision for Civil Engineering in 2025* (Reston, VA: ASCE, 2006).

16. Noble, *America by Design*, xvii.

17. "Graduate Pledge Alliance," accessed September 19, 2011, http://web.mit.edu/mit-cds/gpa/about.html.

18. The other members of the Coordinating Committee are Jens Kabo, Queens University, Canada; Juan Lucena, Colorado School of Mines; Usman Mushtaq; Dean Nieusma, RPI; and Andrés Felipe Valderrama Pineda, co-founder of Engineers Without Borders Colombia."

19. "What This Is About," *Reconstruct: A zine about Engineering, Social Justice, and Peace* 1, 1 (summer 2006): 2.

20. George D. Catalano, *Engineering Ethics: Peace, Justice, and the Earth* (San Rafael, CA: Morgan and Claypool, 2006); Caroline Baillie and George D. Catalano, *Engineering and Society: Working toward Social Justice. Part III, Engineering: Windows on Society* (San Rafael, CA: Morgan and Claypool, 2009).

21. "Welcome," *Reconstruct: Engineering, Social Justice, and Peace (ESJP)* 2 (summer 2009): 2.

22. Engineers Without Borders–USA, accessed September 19, 2011, http://www.ewb-usa.org.

23. Dean Nieusma and Donna Riley, "Designs on Development: Engineering, Globalization, and Social Justice," *Engineering Studies* 2, no. 1 (April 2010): 29–59.

24. Engineers for a Sustainable World, accessed September 19, 2011, http://www.esustainableworld.org.

25. Dean Nieusma, "'Sustainability' as an Integrative Lens for Engineering Education: Initial Reflections on Four Approaches Taken at Rensselaer," paper presented at the ASEE Annual Conference, Austin, TX, 2009; Nicholas Sakellariou, "Engineers and Sustainability: Between Invisibility and (Yet Another) Story of Contextualization in a Period of Crisis," unpublished manuscript, 2010.

26. Jennifer Brown, "Engineering's Soft Side," *Denver Post* (September 30, 2005), A-01.

27. Muñoz quoted in Jeffrey Rubenstone, "Humanitarian Engineering: Not Just Another Charity Case," *Engineering News-Record* (January 1, 2007). http://enr.construction.com/people/ENRNext/archives/070120.asp?&textonly=1.

28. "Graduate Programs for Understanding and Managing Accelerating Change," Acceleration Watch, accessed September 19, 2011, http://www.accelerationwatch.com/degree.html.

29. Ray Kurzweil, *The Singularity Is Near: When Humans Transcend Biology* (New York: Viking, 2005).

30. Peter Diamandis as quoted in NASA Ames Research Center and NASA Research Park in Silicon Valley, *Economic Benefits Study* (Moffett Field, CA: NASA, 2010), 69.

31. Singularity University, *Exponential Technologies Executive Program* (Moffett Field, CA: Singularity University, 2011), 3.

32. National Academy of Engineering, *The Engineer of 2020: Visions of Engineering in the New Century* (Washington, DC: National Academies Press, 2004), 1.

References

Archival Sources

California Institute of Technology Archives, Pasadena, CA
    California Institute of Technology Historical Files
    Committee on the Freshman Year 1964–1967
Claremont Colleges Library, Pomona, CA
    George McKelvey Papers
Getty Research Institute, Los Angeles, CA
    Experiments in Art and Technology Records (940003)
Harvard University Archives, Cambridge, MA
    Records of the Harvard University Program on Technology and Society
MIT Institute Archives and Special Collections, Cambridge, MA
    Ascher Shapiro Papers (MC387)
    Department of Humanities Records (AC404)
    Department of Mechanical Engineering, Records (AC259)
    MIT Program in Science, Technology, and Society (AC363)
    Technology and Culture Seminar Records (AC358)
    Urban Systems Laboratory Records (AC366)
Princeton University Seeley G. Mudd Manuscript Library
    Steve M. Slaby Papers
Swarthmore College Peace Collection, Swarthmore, PA
    Victor Paschkis Papers
UCLA University Archives
    Chancellor's Office Administrative Subject Files of Franklin Murphy (RS401)
University of Utah Library
    Simon Ramo Papers

The Technical Press

Much of engineers' reconciliation with the meaning of technology took place in member society journals, trade and company magazines, and student and underground newspapers. In the over 250 of such works cited, I have relied on certain publications more heavily than others, in some cases reading multi-decade runs. To prevent the bibliography from becoming unwieldy, I have included full bibliographic information for the most vital articles and those from which I have reproduced images. I direct the reader to the endnotes for specific references. Among the most cited are *Chemical Engineering Progress, Civil Engineering, Engineering Education, IEEE Spectrum, Innovation, Mechanical Engineering, Spark, Tech Engineering News,* and *Technology Review.* Additionally I have followed engineers' conversations in *Air Force/Space Digest, American Engineer, American Machinist, Armed Forces Chemical Journal, ASME Transactions, Astronautics and Aeronautics, Aviation Week and Space Technology, Bulletin of the Atomic Scientists, California Tech, Chemical and Engineering News, Chemical Engineering Education, Computer Bulletin, Computers and Automation, Data Link, Electronic News, Electronics, Engineering News Record, Experimental Mechanics, GE Resistor, IEEE CSIT Newsletter, IEEE Technology and Society Magazine, IEEE Transactions on Aerospace and Electronic Systems, Industrial Research, Industrial Water Engineering, Instrumentation Technology, International Science and Technology, Machine Design, Metal Progress, New Scientist, Proceedings of the Institute of Radio Engineers, Product Engineering, Professional Engineer, RCA Engineer, Reconstruct, Science, Scientific American, Signal/Noise, Spartan Engineer, Technos,* and *Yale Scientific.*

Printed and Electronic Sources

Acceleration Watch. "Graduate Programs for Understanding and Managing Accelerating Change." Accessed September 19, 2011. http://www.accelerationwatch.com/degree.html.

Achenbach, Joel. 2008. "The Future Is Now." *Washington Post* (April 13): B1.

Ad hoc Committee on the Triple Revolution. 1964. "The Triple Revolution." *Liberation* 9, no. 2 (April): 9–15.

Ad hoc Faculty Committee on Aims and Goals. 1969. *Aims and Goals of the California Institute of Technology.* Vol. 2: *General Problems of Growth and Change of Caltech,* Vol. 3: *Introducing the Social and Behavioral Sciences at Caltech,* Vol. 4: *The Humanities at Caltech.* Pasadena: California Institute of Technology.

Air Policy Commission. 1948. "Survival in the Air Age." In Pursell, *The Military-Industrial Complex*: 178–97.

Akin, William E. 1977. *Technocracy and the American Dream: The Technocrat Movement, 1900–1941.* Berkeley: University of California Press.

Alder, Ken. 1997. *Engineering the Revolution: Arms and Enlightenment in France, 1763–1815*. Princeton: Princeton University Press.

Aldridge, Bob. 1973. "The Forging of an Engineer's Conscience." *Spark* 3, no. 2 (fall): 2–5.

Alger, Philip. L., N. A. Christensen, and Sterling P. Olmsted, eds. 1965. *Ethical Problems in Engineering*. New York: Wiley.

Allen, Jonathan, ed. 1970. *March 4: Scientists, Students, and Society*. Cambridge: MIT Press.

Allen, Michael Thad, and Gabrielle Hecht, eds. 2001. *Technologies of Power: Essays in Honor of Thomas Parke Hughes and Agatha Chipley Hughes*. Cambridge: MIT Press.

American Society for Engineering Education. 1965. *Goals of Engineering Education: Preliminary Report*. Urbana, IL: ASME.

Anonymous. 1974. *The Mute Engineers*. Princeton, NJ: Literary Publishers.

Arendt, Hannah. 1963. *Eichmann in Jerusalem: A Report on the Banality of Evil*. New York: Viking.

Arthur D. Little, Inc. 1963. *Patterns and Problems of Technical Innovation in American Industry*. Washington, DC: US Government Printing Office.

ASCE Steering Committee to Plan a Summit on the Future of the Civil Engineering Profession in 2025. 2006. *The Vision for Civil Engineering in 2025*. Reston, VA: ASCE.

ASEE Humanistic–Social Research Project. 1968. *Liberal Learning for the Engineer: Report of the ASEE Humanistic–Social Research Project*. Washington, DC: ASEE.

ASME. 1970. *ASME Goals: A Collection of Papers Prepared for the ASME Goals Conference*. New York: ASME.

Associated Students of the California Institute of Technology. 1968. *Air Pollution Project: An Educational Experiment in Self-directed Research*. Pasadena: California Institute of Technology.

Augustine, Dolores L. 2007. *Red Prometheus: Engineering and Dictatorship in East Germany, 1945–1990*. Cambridge: MIT Press.

"Automation House: A Philosophy for Living in a World of Change." *New York Times* (February 1, 1970): AS 2.

Bailes, Kendall E. 1978. *Technology and Society under Lenin and Stalin: Origins of the Soviet Technical Intelligentsia, 1917–1941*. Princeton: Princeton University Press.

Baillie, Caroline, and George D. Catalano. 2009. *Engineering and Society: Working towards Social Justice. Part III, Engineering: Windows on Society*. San Rafael, CA: Morgan and Claypool.

Baker, George Pierce, and American Society of Mechanical Engineers. 1930. *Control: A Pageant of Engineering Progress*. New York: ASME.

Basalla, George. 1973. "Addressing a Central Problem." *Science* 180, no. 4086 (May 11): 582–84.

Bassett, Ross Knox. 2002. *To the Digital Age: Research Labs, Start-up Companies, and the Rise of MOS Technology.* Baltimore: Johns Hopkins University Press.

Baum, Robert J., and Albert Flores, eds. 1978. *Ethical Problems in Engineering.* Troy, NY: Center for the Study of the Human Dimensions of Science and Technology.

Baumgartner, John Stanley. 1970. *The Lonely Warriors: Case for the Military-Industrial Complex.* Los Angeles: Nash.

Beakley, George C., and H. W. Leach. 1972. *Engineering: An Introduction to a Creative Profession,* 2nd ed. New York: Macmillan.

Beakley, George C., and Ernest G. Chilton. 1974. *Design: Serving the Needs of Man.* New York: Macmillan.

Beck, Ulrich. 1992. *Risk Society: Towards a New Modernity.* London: Sage.

Beers, David. 1996. *Blue Sky Dream: A Memoir of America's Fall from Grace.* New York: Doubleday.

Belanger, Dian Olson. 1998. *Enabling American Innovation: Engineering and the National Science Foundation.* West Lafayette, IN: Purdue University Press.

Bell, Daniel. 1960. *The End of Ideology: On the Exhaustion of Political Ideas in the Fifties.* Glencoe, IL: Free Press.

Best, Marshall A. 1963. "In Books, They Call It a Revolution." *Daedalus* 92, no. 1 (winter): 30–41.

Bethlehem Steel. 1970. "Concern." *Tech Engineering News* 52, no. 7 (December): 20.

Bimber, Bruce A. 1996. *The Politics of Expertise in Congress: The Rise and Fall of the Office of Technology Assessment.* Albany: State University of New York Press.

Bix, Amy Sue. 1993. "'Backing Into Sponsored Research': Physics and Engineering at Princeton University, 1945–1970." *History of Higher Education Annual* 13: 9–52.

Bix, Amy Sue. 2000. *Inventing Ourselves out of Jobs? America's Debate over Technological Unemployment, 1929–1981.* Studies in Industry and Society. Baltimore: Johns Hopkins University Press.

Boelter, Llewellyn M. K. 1963. *Education for the Profession.* Los Angeles: UCLA Department of Engineering.

Borgmann, Carl W. 1964. *The Ford Foundation's Role in Engineering Education.* New York: Ford Foundation.

Boulding, Kenneth Ewart. 1964. *The Meaning of the Twentieth Century: The Great Transition.* New York: Harper and Row.

Boyd, John M. 1972. "Science Is Dead—Long Live Technology!" *Engineering Education* 62 (8): 892–95.

Boyer, Paul. 1987. *By the Bomb's Early Light: American Thought and Culture at the Dawn of the Atomic Age*. New York: Pantheon Books.

Brick, Howard. 1992. "Optimism of the Mind: Imagining Post-industrial Society in the 1960s and 1970s." *American Quarterly* 44 (September): 348–80.

Brooks, Harvey. 1967. "Dilemmas of Engineering Education." *IEEE Spectrum* 4, no. 2 (February): 89–91.

Brown, Gordon S. 1959. "The Engineering of Science." *Technology Review* 62, no. 2 (December): 19–22, 48–49.

Brown, Gordon S. 1962. "New Horizons in Engineering Education." *Daedalus* 91, no. 2: 341–61.

Brown, Jennifer. 2005. "Engineering's Soft Side." *Denver Post* (September 30): A-1.

Buckley, William F. Jr. 1969. "What's behind 'March 4—MIT.'" *Los Angeles Times* (February 17): B8.

Buhl, Harold R. 1960. *Creative Engineering Design*. Ames: Iowa State University Press.

Burke, John G., ed. 1966. *The New Technology and Human Values*. Belmont, CA: Wadsworth.

Burnett, D. Graham. 1999. "A View from the Bridge: The Two Cultures Debate, Its Legacy, and the History of Science." *Daedalus* 128, no. 2: 193–218.

Burriss, Stanley W. 1971. "An Industry View." In *Women in Engineering: Bridging the Gap between Society and Technology*, George Bugliarello, Vivian Cardwell, Olive Salembier, and Winifred White, eds. Chicago: University of Illinois at Chicago Circle, 15–27.

Calhoun, Daniel Hovey. 1960. *The American Civil Engineer: Origins and Conflict*. Cambridge, MA: Technology Press.

Calvert, Monte A. 1967. *The Mechanical Engineer in America, 1830–1910*. Baltimore: Johns Hopkins University Press.

Campbell, Bonham, ed. 1963. *Proceedings of the Conference on the Humanities in Engineering Curriculum*. Los Angeles: Department of Engineering, UCLA.

Carson, Rachel. 1962. *Silent Spring*. Boston: Houghton Mifflin.

Catalano, George D. 2006. *Engineering Ethics: Peace, Justice, and the Earth*. San Rafael, CA: Morgan and Claypool.

Center for Policy Alternatives. 1975. *Future Directions for Engineering Education: System Response to a Changing World*. Washington, DC: ASEE.

Chambers, Clarke A. 1958. "The Belief in Progress in Twentieth-Century America." *Journal of the History of Ideas* 19, no. 2: 197–224.

Cohen-Cole, Jamie. 2009. "The Creative American: Cold War Salons, Social Science, and the Cure for Modern Society." *Isis* 100, no. 2 (June): 219–62.

Cohen, Lizabeth. 2003. *A Consumers' Republic: The Politics of Mass Consumption in Postwar America*. New York: Vintage.

Commission on MIT Education. 1971. *Creative Renewal in a Time of Crisis*. Cambridge: MIT.

Committee for Social Responsibility in Engineering. 1971. "Packard???" *Spark* 1, no. 1 (spring): 46–47.

Committee on Evaluation of Engineering Education. 1955. "Summary of the Report on Evaluation of Engineering Education." *Journal of Engineering Education* 46, no. 1 (September): 25–60.

Committee on Science and Astronautics US House of Representatives. 1969. *Technology: Processes of Assessment and Choice, Report of the National Academy of Sciences.* Washington, DC: US Government Printing Office.

Cowan, Ruth Schwartz. 2010. "Looking Back in Order to Move Forward: John McDermott, 'Technology: The Opiate of the Intellectuals.'" *Technology and Culture* 51, no. 1 (January): 199–215.

"Creative Engineering." *Industrial Water Engineering* 5, no. 8 (August 1968): 26–27.

Cutcliffe, Stephen H. 1990. "The STS Curriculum: What Have We Learned in Twenty Years?" *Science, Technology, and Human Values* 15, no. 3 (summer): 360–72.

Cutler Hammer. 1967. "Art at AIL." *IEEE Spectrum* 4, no. 5 (May): 5.

Daniel, James. 1965. "VITA Has the Answer." *Reader's Digest* 87, no. 522 (October): 123–26.

Davenport, William Henry. 1990. *William H. Davenport: Founding Faculty, Chairman of the Humanities and Social Sciences, Willard W. Keit, Jr. Fellow in Humanities, 1957–1973: Oral History Interview.* Interview by Enid Hart Douglass. Claremont College Library.

Davenport, William Henry. 1970. *The One Culture: Unified Engineering Series.* New York: Pergamon Press.

Davenport, William Henry, and Daniel M. Rosenthal. 1967. *Engineering: Its Role and Function in Human Society.* New York: Pergamon Press.

Davenport, William Henry, and J. P. Frankel. 1968. *The Applied Humanities.* Los Angeles: Department of Engineering, UCLA.

Davis, Douglas. 1973. *Art and the Future: A History-Prophecy of the Collaboration Between Science, Technology and Art.* New York: Praeger.

Diebold, John. 1952. *Automation: The Advent of the Automatic Factory.* New York: Van Nostrand.

de Nevers, Noel. 1972. *Technology and Society.* Reading, MA: Addison-Wesley.

Dennison, Edward S. 1971. "Technology Transfer within Developing Areas." In *Metrology and Standardization in Less Developed Countries: The Role of a National Capability for Industrializing Economies.* H. L. Mason and H. S. Peiser, eds. Washington, DC: National Bureau of Standards, 293–304.

DeSimone, Daniel V., ed. 1968. *Education for Innovation.* Oxford: Pergamon Press.

Deutsch and Shea, Inc. 1959. *Company Climate and Creativity: 105 Outstanding Authorities Present Their Views.* New York: Industrial Relations News.

Dickson, David. 1979. "MIT Puts Some Polish on a Tarnished Image." *Nature* 281 (October 25): 627–28.

Dorf, Richard C. 1968. "Technology and Man's Future: University of Santa Clara April 1968." *Technology and Culture* 9, no. 4 (October): 580–83.

Dorf, Richard C. 1972. *Technology and Society.* San Francisco: Boyd and Fraser.

Downey, Gary Lee. 1998. *The Machine in Me: An Anthropologist Sits among Computer Engineers.* New York: Routledge.

Downey, Gary Lee. 2009. "What Is Engineering Studies For? Dominant Practices and Scalable Scholarship." *Engineering Studies* 1, no. 1: 55–76.

Downey, Gary Lee, and Juan C. Lucena. 2004. "Knowledge and Professional Identity in Engineering: Code-Switching and the Metrics of Progress." *History and Technology* 20, no. 4 (December): 393–420.

Drucker, Peter. 1959. *Landmarks of Tomorrow.* New York: Harper.

Durbin, Paul T. 1997. "Engineering Ethics and Social Responsibility: Reflections on Recent Developments in the USA." *Bulletin of Science, Technology and Society* 17, no. 2/3: 77–83.

Durbin, Paul T. 1992. *Social Responsibility in Science, Technology, and Medicine.* Bethlehem, PA: Lehigh University Press.

EDP Humanities Subcommittee, ed. 1963. *Studies of Courses and Sequences in Humanities, Fine Arts and Social Sciences for Engineering Students.* Los Angeles: Department of Engineering, UCLA.

Egan, Michael. 2007. *Barry Commoner and the Science of Survival: The Remaking of American Environmentalism.* Cambridge: MIT Press.

Ellul, Jacques. 1964. *The Technological Society.* Wilkinson, John, trans. New York: Knopf.

Engineering Concepts Curriculum Project. 1969. *The Man Made World: A Course on the Theories and Techniques that Contribute to our Technological Civilization*, vol. 1. New York: McGraw-Hill.

Engineering Concepts Curriculum Project. 1973. *Man and His Technology.* New York: McGraw-Hill.

Engineering Concepts Curriculum Project. 1964. *Working Papers of the Summer Study Report on the Engineering Concepts Curriculum Project.* Washington, DC: Commission on Engineering Education.

Engineering Manpower Commission. 1963. *Engineering Student Attrition: Is It Undermining Our Nation's Engineering Manpower?* New York: EJC.

Engineers Joint Council. 1962. *The Nation's Engineering Research Needs, 1965–1985.* New York: EJC.

Engineers Joint Council. 1964a. *Engineering Manpower in Profile.* New York: EJC.

Engineers Joint Council. 1964b. *National Engineering Problems: Summary Report on Symposium Sponsored by Engineers Joint Council, January 12–14, 1964.* New York: EJC.

Engineers Joint Council. 1971. *A Profile of the Engineering Profession: A Report from the 1969 National Engineers Register.* New York: EJC.

Engineers Without Borders-USA. 2010. *Strategic Plan.* Boulder, CO: EWB-USA.

Etzioni, Amitai. 1972. "Minerva: An Electronic Town Hall." *Policy Sciences 3,* no. 4: 457–74.

Ewing, David W., ed. 1970. *Technological Change and Management: The John Diebold Lectures, 1968–1970.* Boston: Harvard University Graduate School of Business Administration.

Fabun, Don. 1969. *The Children of Change.* Beverly Hills, CA: Glencoe Press.

Fay, James A. 1971. *Physical Processes in the Spread of Oil on a Water Surface.* Springfield, VA: National Technical Information Service.

Fay, James A., and D. Golomb. 2002. *Energy and the Environment. MIT-Pappalardo Series in Mechanical Engineering.* New York: Oxford University Press.

Florman, Samuel C. 1968a. *Adventure in the Arts: An Engineer's Guide to Great Books.* New York: McGraw-Hill.

Florman, Samuel C. 1968b. *Engineering and the Liberal Arts: A Technologist's Guide to History, Literature, Philosophy, Art, and Music, Mcgraw-Hill Series in Continuing Education for Engineers.* New York: McGraw-Hill.

Florman, Samuel C. 1972. "Anti-technology: The New Myth." *Civil Engineering 42,* no. 1 (January): 68–71.

Florman, Samuel C. 1974. "Creativity and the Anti-Technologists." In *Civil Engineers in the World around Us,* M. D. Morris, ed. New York: American Society of Civil Engineers.

Florman, Samuel C. 1976. *The Existential Pleasures of Engineering.* New York: St. Martin's Press.

Florman, Samuel C. 1981a. *Blaming Technology: The Irrational Search for Scapegoats.* New York: St. Martin's Press.

Florman, Samuel C. 1981b. "Do It Yourself Is the Message." *New York Times Book Review* (January 25): 3, 24.

Appropriate Technology: Hearings. 1978. *Day 1, Before the Subcommittee on Domestic and International Scientific Planning, Analysis, and Cooperation of the Committee on Science and Technology, US House of Representatives.* 95th Cong.: 135–42 (Statement of Thomas H. Fox).

Friedman, Thomas. 2005. "It's a Flat World After All." *New York Times Magazine* (April 3): 33–77.

Fries, Sylvia Doughty. 1983. "Expertise against Politics: Technology as Ideology on Capital Hill, 1966–1972." *Science, Technology and Human Values* 8, no. 2: 6–15.

Fulbright, J. William. 1967. "The Great Society Is a Sick Society." *New York Times Magazine* (August 20): 30, 88–96.

Fuller, R. Buckminster. 1970. "Commitment to Humanity." *Humanist* 30 (May/June): 28–33.

Furnas, C. C., and Joe McCarthy. 1966. *The Engineer*. Life Science Library. New York: Time, Inc.

Galbraith, John Kenneth. 1967. *The New Industrial State*. Boston: Houghton Mifflin.

Galison, Peter. 1998. "The Americanization of Unity." *Daedalus* 127, no. 1: 45–71.

Geiger, Roger L. 1993. *Research and Relevant Knowledge: American Research Universities since World War II*. New York: Oxford University Press.

General Electric Company. 1968. "Dan Johnson Has a Flair for Making Things." *Spartan Engineer* 21, no. 4 (May): 26.

Gilbert, G. M. 1947. *Nuremberg Diary*. New York: Farrar, Strauss, and Giroux.

Ginzberg, Carlo. 1980. *The Cheese and the Worms: The Cosmos of a Sixteenth-Century Miller*, John Tedeschi and Anne Teceschi, trans. Baltimore: Johns Hopkins University Press.

"Goals: Basis for Action Programs, 'Making Technology a True Servant of Man.'" *Mechanical Engineering* 93, no. 4 (April 1971): 16–19.

Goals Report Committee. 1970. "Goals: A Proposed Statement." *Mechanical Engineering* 92, no. 4 (April): 18–23.

Goldhoff, Robert M. 1970. "VITA and Materials Technology: Coupling Technical Skills with Problems of World Development." In *An Interamerican Approach for the Seventies*, vol. 2: *Materials Education*, ed. by Southwest Research Institute. New York: ASME, 27–33.

Goodman, Paul. 1969a. "The Case against Technology." *Innovation* 2 (June): 36–47.

Goodman, Paul. 1969b. "Can Technology Be Humane?" *New York Review of Books* 13, no. 9 (November 20): 27–34.

Goodyear, Anne Collins. 2004. "Gyorgy Kepes, Billy Klüver, and American Art of the 1960s: Defining Attitudes toward Science and Technology." *Science in Context* 17, no. 4: 611–35.

Goodyear, Anne Collins. 2008. "From Technophilia to Technophobia: The Impact of the Vietnam War on the Reception of 'Art and Technology.'" *Leonardo* 41, no. 2 (April): 169–73.

Gould, Heywood. 1971. *Corporation Freak*. New York: Tower Publications.

Gouldner, Alvin W. 1976. *The Dialectic of Ideology and Technology: The Origins, Grammar, and Future of Ideology*. New York: Seabury Press.

"Graduate Pledge Alliance." Accessed September 19, 2011. http://web.mit.edu/mit-cds/gpa/about.html.

Great Atlantic and Pacific School Conspiracy. 1972. *Doing Your Own School: A Practical Guide to Starting and Operating a Community School.* Boston: Beacon Press.

Gross, Alex. 1970. "Who Is Being Eaten?" *East Village Other* (March 3): 14.

Gustafson, James M., and James T. Laney, eds. 1968. *On Being Responsible: Issues in Personal Ethics.* New York: Harper and Row.

Hamilton Standard. 1966. "Wanted: An Engineer Who Wants to Change Things." *Mechanical Engineering* 88, no. 6 (June): 145.

Harvard University Program on Technology and Society. 1965. *First Annual Report of the Executive Director.* Cambridge: Harvard University Program on Technology and Society.

Harvard University Program on Technology and Society. 1967. *Third Annual Report of the Executive Director.* Cambridge: Harvard University Program on Technology and Society.

Harvard University Program on Technology and Society. 1972. *A Final Review, 1964–1972.* Cambridge: Harvard University Press.

Hayes, Wallace D., and Ronald F. Probstein. 1959. *Hypersonic Flow Theory.* New York: Academic Press.

Heilbroner, Robert L. 1960. *The Future as History: The Historic Currents of Our Time and the Direction in Which They Are Taking America.* New York: Harper.

Henthorn, Cynthia Lee. 2006. *From Submarines to Suburbs: Selling a Better America, 1939–1959.* Athens: Ohio University Press.

Herf, Jeffrey. 1984. *Reactionary Modernism: Technology, Culture, and Politics in Weimar and the Third Reich.* New York: Cambridge University Press.

Hoffman, Harold L. 1968. "VITA—A Sociotechnical Challenge for the IEEE." *IEEE Spectrum* 5 no. 8 (August): 82–89.

Hollomon, J. Herbert. 1968. "Creative Engineering and the Needs of Society." In DeSimone, *Education for Innovation*, 23–30.

Hollomon, J. Herbert 1972. "Technology Policy: The U.S. Picks Its Way." *IEEE Spectrum* 9, no. 8 (August): 72–81.

Hollomon, J. Herbert. 1991. "Engineering's Great Challenge—The 1960s." In *Engineering as a Social Enterprise*, Hedy E. Sladovich, ed. Washington, DC: National Academy Press, 104–10.

Horgan, J. D. 1973. "Technology and Human Values: The 'Circle of Action.'" *Mechanical Engineering* 95, no. 8 (August): 19–22.

Horosz, William. 1975. *The Crisis of Responsibility: Man as the Source of Accountability.* Norman: University of Oklahoma Press.

Hughes, Thomas P., ed. 1975. *Changing Attitudes toward American Technology.* New York: Harper and Row.

Hughes, Thomas P. 1989. *American Genesis: A Century of Invention and Technological Enthusiasm, 1870–1970.* New York: Viking.

Hughes, Thomas P. 1998. *Rescuing Prometheus*. New York: Pantheon Books.

Humanistic–Social Research Project. 1956. *General Education in Engineering*. Urbana, IL: ASEE.

Hunter, James Davidson. 1991. *Culture Wars: The Struggle to Define America*. New York: Basic Books.

"The Innovation Group!" *Wall Street Journal* (April 25, 1969): 4.

Institute of Electrical and Electronics Engineers. 1971. *1971 IEEE International Convention Digest*. New York: IEEE.

Interim-Committee on the Social Aspects of Science. 1956. Society in the Scientific Revolution. *Science* 124, no. 3234 (December 21): 1231.

Isaac, Jeffrey C. 1992. *Arendt, Camus, and Modern Rebellion*. New Haven: Yale University Press.

Johnson, Jill. 1966. "Post Mortem." *Village Voice* (December 15): back cover.

Jones, Brendan. 1969. "Developing Lands Call on 'Brain Bank.'" *New York Times* (July 6): F12.

Jones, Caroline A. 1996. *Machine in the Studio: Constructing the Postwar American Artist*. Chicago: University of Chicago Press.

Kaiser, David. 2004. "The Postwar Suburbanization of American Physics." *American Quarterly* 56, no. 4 (December): 851–88.

Kaiser, David. 2011. *How the Hippies Saved Physics: Science, Counterculture, and the Quantum Revival*. New York: Norton.

Kapstein, Ethan Barnaby. 1981. "The Solar Cooker." *Technology and Culture* 22, no. 1 (January): 112–21.

Kargon, Robert, and Stuart W. Leslie. 1994. "Imagined Geographies: Princeton, Stanford and the Boundaries of Useful Knowledge in Postwar America." *Minerva* 32, no. 2 (June): 121–43.

Kargon, Robert, Stuart W. Leslie, and Erica Schoenberger. 1992. "Far Beyond Big Science: Science Regions and the Organization of Research and Development." In *Big Science: The Growth of Large-Scale Research*, Peter Galison and Bruce Hevly, eds. Stanford: Stanford University Press, 334–54.

Kaufman, Harold G., ed. 1975. *Career Management: A Guide to Combating Obsolescence*. New York: IEEE.

Kelves, Daniel J. 1979. *The Physicists: The History of a Scientific Community in Modern America*. New York: Vintage Books.

Kemper, John Dustin. 1967. *The Engineer and His Profession*. New York: Holt, Rinehart and Winston.

Kemper, John Dustin. 1975. *The Engineer and His Profession*, 2nd ed. New York: Holt, Rinehart and Winston.

Kerr, Clark. 1963. *The Uses of the University*. Cambridge: Harvard University Press.

King, Martin Luther Jr. 1963. *Strength to Love*. New York: Harper and Row.

Kline, Ronald R. 1995. "Constructing 'Technology' as 'Applied Science': Public Rhetoric of Scientists and Engineers in the United States, 1880–1945." *Isis* 86, no. 2 (June): 194–221.

Kline, Ronald R. 2006. "Cybernetics, Management Science, and Technology Policy: The Emergence of 'Information Technology' as a Keyword, 1948–1985." *Technology and Culture* 47, no. 3 (July): 513–35.

Kline, Ronald. 2008. "From Progressivism to Engineering Studies: Edwin T. Layton's *The Revolt of the Engineers.*" *Technology and Culture* 49, no. 4 (October): 1018–24.

Klüver, Billy. 1960. "Fragment on Man and the System." In *The Hasty Papers: A One Shot Review.* New York: Alfred Leslie, 45.

Klüver, Billy. 1968. "The Garden Party." In *The Machine: As Seen At the End of the Mechanical Age*, Karl Gunnar Pontus Hulten, ed. New York: Museum of Modern Art, 169–71.

Klüver, Billy, Julie Martin, and Barbara Rose, eds. 1972. *Pavilion.* New York: Dutton.

Knepler, Henry. 1973. "Engineering Education and the Humanities in America." *Leonardo* 6, no. 4: 305–309.

Kozol, Jonathan. 1972. *Free Schools.* Boston: Houghton Mifflin.

Kranzberg, Melvin. 1964. "Technology and Human Values." *Virginia Quarterly Review* 40 (autumn): 578–92.

Kranzberg, Melvin, and Carroll W. Pursell Jr., eds. 1967. "Technology's Challenge." In *Technology in Western Civilization*, vol. 2. New York: Oxford University Press, 695–708.

Kurzweil, Ray. 2005. *The Singularity Is Near: When Humans Transcend Biology.* New York: Viking.

Ladd, Everett Carll Jr., and Seymour Martin Lipset. 1972. Politics of Academic Natural Scientists and Engineers. *Science* 176, no. 4039 (June 9): 1091–1100.

Lauda, Donald P., and Robert D. Ryan. 1971. *Advancing Technology: Its Impact on Society.* Debuque, IA: WC Brown.

Lavin, Chad. 2008. *The Politics of Responsibility.* Urbana: University of Illinois Press.

Lavoie, Francis J. 1972. "The Activist Engineer: Look Who's Getting Involved." *Machine Design* (November 2): 82–88.

Layton, Edwin T. Jr. 1962. "Veblen and the Engineers." *American Quarterly* 14, no. 1 (spring): 64–72.

Layton, Edwin T. Jr. 1986. *The Revolt of the Engineers: Social Responsibility and the American Engineering Profession.* Baltimore: Johns Hopkins University Press.

Lécuyer, Christophe. 2006. *Making Silicon Valley: Innovation and the Growth of High Tech, 1930–1970.* Cambridge: MIT Press.

Ledbetter, James. 2011. *Unwarranted Influence: Dwight D. Eisenhower and the Military-Industrial Complex.* New Haven: Yale University Press.

Lerner, Joshua. 1984. *Volunteers in Technical Assistance: The First Twenty-Five Years of VITA.* St. Louis: Center for Development Technology, Washington University.

Leslie, Stuart W. 1993. *The Cold War and American Science: The Military-Industrial-Academic Complex at MIT and Stanford.* New York: Columbia University Press.

Leventman, Paula Goldman. 1981. *Professionals Out of Work.* New York: Free Press.

Light, Jennifer S. 2003. *From Warfare to Welfare: Defense Intellectuals and Urban Problems in Cold War America.* Baltimore: Johns Hopkins University Press.

Lindgren, Nilo. 1969a. "Art and Technology: I. Steps toward a New Syngerism." *IEEE Spectrum* 6, no. 4 (April): 59–68.

Lindgren, Nilo. 1969b. "Art and Technology: II. A Call for Collaboration." *IEEE Spectrum* 6, no. 5 (May): 46–56.

Lord, M. G. 2005. *Astro Turf: The Private Life of Rocket Science.* New York: Walker.

Lucena, Juan C. 2005. *Defending the Nation: U.S. Policymaking to Create Scientists and Engineers from Sputnik to the 'War against Terrorism.* Lantham, MD: University Press of America.

Maass, William G. 1962. "New Information Services from a Not-So-Old Publishing House." *Journal of Chemical Documentation* 2, no. 1 (January): 46–48.

Macdonald, Dwight. 1957. *The Responsibility of Peoples: And Other Essays in Political Criticism.* London: Gollancz.

Mack, Pamela E. 2001. "What Difference Has Feminism Made to Engineering in the Twentieth Century?" In *Feminism in Twentieth-Century Science, Technology, and Medicine,* Angela N. H. Creager, Elizabeth Lunbeck, and Londa L. Schiebinger, eds. Women in Culture and Society Series. Chicago: University of Chicago Press, 149–68.

Mahoney, Michael S. 2001. "'Engineering *Plus*': Technical Education and the Liberal Arts at Princeton." In *From College to University: Essays on the History of Princeton University,* Anthony T. Grafton and John M. Murrin, eds.http://www.princeton.edu/~hos/Mahoney/seashist/engineer.htm.

Marchand, Roland. 1998. *Creating the Corporate Soul: The Rise of Public Relations and Corporate Imagery in American Big Business.* Berkeley: University of California Press.

Marcuse, Herbert. 1964. *One Dimensional Man: Studies in the Ideology of Advanced Industrial Society.* Boston: Beacon Press.

Marine, Gene. 1969. *America the Raped: The Engineering Mentality and the Devastation of a Continent.* New York: Simon and Schuster.

Markoff, John. 2005. *What the Dormouse Said: How the Sixties Counterculture Shaped the Personal Computer Industry*. New York: Viking.

Marlowe, Donald E. 1964a. "The Legacy of Merlin." *Mechanical Engineering* 86, no. 2 (February): 26–29.

Marlowe, Donald E. 1964b. "The Specter of Socialized Engineering." *Mechanical Engineering* 86, no. 6 (June): 24–25.

Marlowe, Donald E. 1969a. "Public Interest—First Priority in Engineering Design?" *Professional Engineer* 39, no. 2 (February): 23–25.

Marlowe, Donald E. 1969b. "Technology and Society, Part 1: The Public Interest." *Mechanical Engineering* 91, no. 4 (April): 24–26.

Marlowe, Donald E. 1969c. "As the President Sees It." *Mechanical Engineering* 91, no. 9 (September): 81.

Marlowe, Donald E. 1969d. "Prometheus Unbound." *Mechanical Engineering* 91, no. 11 (November): 28–30.

Marlowe, Donald E. 1970a. "The New Luddites." *Mechanical Engineering* 92, no. 3 (March): 12–13.

Marlowe, Donald E. 1970b. "Goals for a More Vital Society." *Mechanical Engineering* 92, no. 4 (April): 17.

Marlowe, Donald E. 1971. "Review: *The Revolt of the Engineers*." *Mechanical Engineering* 93, no. 10 (October): 66.

Martin, Julie. 1967a. "Program." *E.A.T. News* 1, no. 3 (November): 6.

Martin, Julie. 1967b. *E.A.T. News* 1, no. 4 (December 20): 3.

Marx, Leo. 1994. "The Idea of 'Technology' and Postmodern Pessimism." In *Does Technology Drive History?* Smith and Marx, eds.: 238–57.

Marx, Leo. 1997. "Technology: The Emergence of a Hazardous Concept." *Social Research* 64, no. 3 (fall): 965–88.

Massachusetts Institute of Technology. 1969. *Review Panel on Special Laboratories: Final Report*. Cambridge: MIT.

Massachusetts Institute of Technology Committee on Educational Survey. 1949. *Report of the Committee on Educational Survey to the Faculty of the Massachusetts Insitute of Technology*. Cambridge: Technology Press.

McAllister, Don. 1971. "Letter to the Editor." *Aviation Week and Space Technology* 95 (April 5): 50.

McDermott, John. 1969. "Technology: The Opiate of the Intellectuals." *New York Review of Books* (July 31): 25–36.

McDougall, Walter. 1985. *The Heavens and the Earth: A Political History of the Space Age*. New York: Basic Books.

McEnany, John Merritt. 1972. "The Princeton Strike of 1970: A Narrative and Commentary." Senior thesis. Princeton University.

McGee, David. 1995. "Making up Mind: The Early Sociology of Invention." *Technology and Culture* 36, no. 4 (October): 773–801.

McLuhan, Marshall. 1964. *Understanding Media.* New York: McGraw-Hill.

McMahon, A. Michal. 1984. *The Making of a Profession: A Century of Electrical Engineering in America.* New York: IEEE.

McMahon, Darrin M. 2001. *Enemies of the Enlightenment: The French Counter-Enlightenment and the Making of Modernity.* Oxford: Oxford University Press.

Mead, Margaret, ed. 1958. *Cultural Patterns and Technical Change.* New York: New American Library and UN Educational, Scientific and Cultural Organization.

Meiksins, Peter. 1988. "The *Revolt of the Engineers* Reconsidered." *Technology and Culture* 29, no. 2 (April): 219–46.

Meiksins, Peter. 1996a. "Engineers in the United States: A House Divided." In *Engineering Labour: Technical Workers in Comparative Perspective,* Peter Meiksins and Chris Smith, eds. London: Verso, 61–97.

Meiksins, Peter, and Chris Smith, eds. 1996b. *Engineering Labour: Technical Workers in Comparative Perspective.* London: Verso.

Melman, Seymour. 1970. *Pentagon Capitalism: The Political Economy of War.* New York: McGraw-Hill.

Mendelsohn, Everett. 1995. "The Politics of Pessimism: Science and Technology Circa 1968." In *Technology, Pessimism, and Postmodernism,* Yaron Ezrahi, Everett Mendelsohn, and Howard P. Segal, ed. Amherst: University of Massachusetts Press, 175–216.

Merton, Robert K. 1947. "The Machine, the Worker, and the Engineer." *Science* 105, no. 2717 (January 24): 79–84.

Merton, Robert K. 1964. "Foreword" to *The Technological Society* by Jacques Ellul: v–viii.

Mesthene, Emmanuel G. 1962. *The Titanium Decade.* Santa Monica, CA: RAND Corporation.

Mesthene, Emmanuel G. 1964. "Can Only Scientists Make Government Science Policy?" *Science* 145, no. 3629 (July 17): 237–40.

Mesthene, Emmanuel G. 1967a. "The Impacts of Science on Public Policy." *Public Administration Review* 27, no. 2: 97–104.

Mesthene, Emmanuel G. 1967b. "Technology and Wisdom." In *Technology and Social Change,* Emmanuel G. Mesthene, ed. Indianapolis: Bobbs-Merrill, 57–62.

Mesthene, Emmanuel G. 1968. "How Technology Will Shape the Future." *Science* 161, no. 3837: 135–43.

Mesthene, Emmanuel G. 1969a. "Some General Implications of the Research of the Harvard University Program on Technology and Society." *Technology and Culture* 10, no. 4: 489–513.

Mesthene, Emmanuel G. 1969b. "Technology and Human Values." *Science Journal* 5A, no. 4 (October): 45–50.

Mesthene, Emmanuel G. 1969c. *Technology Assessment: Hearings before the Subcommittee on Science, Research, and Development of the Committee on Science and Astronautics, U.S. House of Representatives*, 91st Cong., 13. (Statement of Emmanuel G. Mesthene, director Harvard University Program on Technology and Society).

Mesthene, Emmanuel G., and Herbert Hollomon. 1972. "The New Meaning of Social Responsibility." *Innovation* 28 (February): 2–9.

Mesthene, Emmanuel G., and Organization for Economic Cooperation and Development. 1965. *Ministers Talk about Science: A Summary and Review*. Paris: OECD.

Miller, Paul. 1998. "The Engineer as Catalyst: Billy Klüver on Working with Artists." *IEEE Spectrum* 35, no. 7 (July): 20–29.

Mills, C. Wright. 1951. *White Collar: The American Middle Classes*. New York: Oxford University Press.

Mills, C. Wright. 1956. *The Power Elite*. New York: Oxford University Press.

"The Misuse of Science." *Nation* (February 24, 1969): 228.

Mock, Jesse, ed. 1969. *The Engineer's Responsibility to Society*. New York: ASME.

Montana, Ilene, ed. 1967. *Technology and Culture in Perspective*. Cambridge, MA: Church Society for College Work.

"The Moon and 'Middle America.'" *Time* (August 1, 1969): 10–11.

Moore, Kelly. 2008. *Disrupting Science: Social Movements, American Scientists, and the Politics of the Military, 1945–1975*. Princeton: Princeton University Press.

Morse, Richard. 1972. *Responding to Technical Information Needs in Developing Countries: An Evaluative Review of the VITA International Inquiry Service*. Washington, DC: AID.

Morton, Jack A. 1964. "From Research to Technology." *International Science and Technology* 29 (May): 82–92.

Morton, Jack A. 1966. "The Microelectronics Dilemma." *International Science and Technology* 55 (July): 35–44.

Mumford, Lewis. 1967. *The Myth of the Machine: Technics and Human Development*. New York: Harcourt Brace and World.

Mumford, Lewis. 1970. *The Myth of the Machine: The Pentagon of Power*. New York: Harcourt Brace Jovanovich.

Murray, Chris. 1974. "From GE to the Syracuse Peace Council." *Spark* 4, no.1 (spring): 2–3.

Nader, Ralph. 1965. *Unsafe at Any Speed: The Designed-in Dangers of the American Automobile*. New York: Grossman.

NASA Ames Research Center and NASA Research Park in Silicon Valley. 2010. *Economic Benefits Study*. Mofett Field, CA: NASA.

National Academy of Engineering. 1969. *A Study of Technology Assessment: Report of the Committee on Public Engineering Policy*. Washington, DC: US Government Printing Office.

National Academy of Engineering. 2004. *The Engineer of 2020: Visions of Engineering in the New Century*. Washington, DC: National Academies Press.

National Industrial Conference Board, ed. 1966. *The Challenge of Technology: Linking Business, Science, and the Humanities in Examining Management and Man in the Computer Age*. New York: The Conference Board.

National Research Council Panel on Engineering Interactions with Society. 1985. *Engineering Education and Practice in the United States: Engineering in Society*. Washington, DC: National Academy Press.

National Science Foundation. 1964. *Scientific and Technical Manpower Resources: Summary Information on Employment, Characteristics, Supply, and Training*. Washington, DC: US Government Printing Office.

National Science Foundation. 1967. *Geographic Distribution of Federal Funds for Research and Development: Fiscal Year 1965*. Washington, DC: US Government Printing Office.

National Science Foundation. 1968. *Employment of Scientists and Engineers in the United States, 1950–1966*. Washington, DC: US Government Printing Office.

National Science Foundation. 1972. *Unemployment Rates and Employment Characteristics for Scientists and Engineers, 1971*. Washington, DC: US Government Printing Office.

National Science Foundation. 1982. *Women and Minorities in Science and Engineering*. Washington, DC: NSF.

National Technology Foundation Act. 1980. *Hearings on H.R. 6910, before the Subcommittee on Science, Research and Technology of the Committee on Science and Technology*, 96th Cong.

Nelkin, Dorothy. 1971. *The Politics of Housing Innovation: The Fate of the Civilian Industrial Technology Program*. Ithaca: Cornell University Press.

Nelkin, Dorothy. 1972. *The University and Military Research: Moral Politics at MIT*. Ithaca: Cornell University Press.

Newcombe, George M. 1969. "Engineering, a Modern Profession in a Moral Society—A Student's Viewpoint." In *Are Engineering and Science Relevant to Moral Issues in a Technological Society?* New York: EJC, 7–11.

Niebuhr, H. Richard. 1963. *The Responsible Self: An Essay in Christian Moral Philosophy*. New York: Harper and Row.

Nieusma, Dean. 2009. "'Sustainability' as an Integrative Lens for Engineering Education: Initial Reflections on Four Approaches Taken at Rensselaer." Paper presented at the ASEE Annual Conference, Austin, TX.

Nieusma, Dean, and Donna Riley. 2010. "Designs on Development: Engineering, Globalization, and Social Justice." *Engineering Studies* 2, no. 1 (April): 29–59.

Noble, Daniel E. 1970. *Noble Comments*. Phoenix, AZ: Motorola Inc.

Noble, David F. 1977. *American By Design: Science, Technology, and the Rise of Corporate Capitalism*. New York: Knopf.

Norgen, Paul Herbert, and Aaron W. Warner. 1966. *Obsolescence and Updating of Engineers' and Scientists' Skills: Final Revised Report*. New York: Columbia

University Seminar on Technology and Social Change, and US Office of Manpower Policy Evaluation and Research.

Nourse, Alan E. 1962. *So You Want to Be an Engineer.* New York: Harper and Row.

O'Connell, James D., Eugene G. Fubini, Kenneth G. McKay, James Hillier, and J. Herbert Hollomon. 1969. "Electronically Expanding the Citizen's World." *IEEE Spectrum* 6, no. 7 (July): 30–40.

Ogburn, William Fielding. 1922. *Social Change with Respect to Culture and Original Nature.* New York: Huebsch.

Ogburn, William Fielding. 1933. "Laggard Parts of Our Social Machine." *New York Times Magazine* (April 16): 5, 19.

Ogburn, William Fielding. 1934. *You and Machines.* Chicago: University of Chicago Press.

Ogburn, William Fielding. 1936. Technology and Governmental Change. *Journal of Business of the University of Chicago* 9, no. 1: 1–13.

Oldenziel, Ruth. 1999. *Making Technology Masculine: Men, Women, and Modern Machines in America, 1870–1945.* Amsterdam: Amsterdam University Press.

Oldenziel, Ruth. 2006. "Signifying Semantics for a History of Technology." *Technology and Culture* 47, no. 3 (July): 477–85.

Olin Corporation. 1969. "Will Olin Turn You into an Organization Robot?" *Journal of College Placement* 30, no. 1 (October/November):83.

Ortolano, Guy. 2009. *The Two Cultures Controversy: Science, Literature and Cultural Politics in Postwar Britain.* Cambridge: Cambridge University Press.

Parisi, A. J. 1970. "New Kind of Conference Focuses on New Ideas in Technology." *Product Engineering* (March 2): 22–24.

Paschkis, Victor. 1947a. "Double Standards." *Friends Intelligencer* (August 30): 463.

Paschkis, Victor. 1947b. "The Heat and Mass Flow Analyzer Laboratory." *Metal Progress* 52, no. 6 (November): 813–18.

Paschkis, Victor. 1964. "To the Editor." *Mechanical Engineering* 86, no. 8 (August): 74.

Paschkis, Victor. 1965. "The Disease They-Mindedness and Its Cure." *Journal of Human Relations* 13, no. 2: 178–84.

Paschkis, Victor. 1966. "Cybernation and Civil Rights." In *The Evolving Society: The Proceedings of the First Annual Conference on the Cybercultural Revolution— Cybernetics and Automation,* Alice Mary Hilton and Institute for Cybercultural Research, eds. New York: Institute for Cybercultural Research, 357–64.

Paschkis, Victor. 1970a. "To the Editor." *Mechanical Engineering* 92, no. 6 (June): 81.

Paschkis, Victor. 1970b. "Assessment—By Whom, for Whom?" ASME Winter Annual Meeting, 70-WA/Av-5.

Paschkis, Victor. 1971. "Moving toward Responsible Technology." ASME Winter Annual Meeting, 71-WA/Av-1.

Pearson, Donald S. 1958. *Creativeness for Engineers: A Philosophy and a Practice*. University Park: Pennsylvania State University.

Perlstein, Rick. 2008. *Nixonland: The Rise of a President and the Fracturing of America*. New York: Simon and Schuster.

Perrucci, Robert, and Joel Emery Gerstl, eds. 1969a. *The Engineers and the Social System*. New York: Wiley.

Perrucci, Robert, and Joel Emery Gerstl, eds. 1969b. *Profession without Community: Engineers in American Society*. New York: Random House.

Petroski, Henry. 1985. *To Engineer Is Human: The Role of Failure in Successful Design*. New York: St. Martin's Press.

Pilisuk, Marc, and Thomas Hayden. 1972. "Is There a Military-Industrial Complex?" Reprinted in Pursell, *Military-Industrial Complex*: 51–80.

Platt, Joseph B. 1994. *Harvey Mudd College: The First Twenty Years*. Santa Barbara, CA: Fithian Press.

Prelinger, Megan. 2010. *Another Science Fiction: Advertising the Space Race, 1957–1962*. New York: Blast Books.

Price, Don K. 1965. *The Scientific Estate*. Cambridge: Belknap Press of Harvard University Press.

Probstein, Ronald F. 1969. "Reconversion and Non-military Research Opportunities." *Astronautics and Aeronautics* 7, no. 10 (October): 50–56.

Probstein, Ronald F. 1970. "Application of Aerospace and Defense Industry Technology to Environmental Problems." *Hearings before the Subcommittee on Government Operations House of Representatives*, 91st Cong.: 74–77.

Probstein, Ronald F. 1971. "Taking Effective Federal Action on the Environment." *Astronautics and Aeronautics* 9, no. 1 (January): 15, 80.

Pursell, Carroll W. Jr., ed. 1972. *The Military-Industrial Complex*. New York: Harper and Row.

Pursell, Carroll W. Jr., ed. 1979. "The History of Technology as a Source of Appropriate Technology." *Public Historian* 1, no. 2 (winter): 15–22.

Pursell, Carroll W. Jr., ed. 1993. "The Rise and Fall of the Appropriate Technology Movement in the United States, 1965–1985." *Technology and Culture* 34, no. 3 (July): 629–37.

Rae, John. 1975. 'Engineers Are People." *Technology and Culture* 16, no. 3 (July): 404–18.

Ramo, Simon. 1956. "Weapons Systems Engineering and Changing Role of the Scientist." *Armed Forces Chemical Journal* (July/August): 23.

Ramo, Simon. 1957. "The Impact of Systems Engineering on Education." *American Engineer* 27, no. 10 (October): 11–15.

Ramo, Simon. 1962. "The Impact of Missiles and Space on Electronics." *Proceedings of the Institute of Radio Engineers* 50, no. 5 (May): 1237–41.

Ramo, Simon. 1965. "The Coming Technological Society." In *NATO's Fifteen Nations* (December 1964/January): 66–73.

Ramo, Simon. 1969a. "Comment: The Anticipation of Change." *Technology and Culture* 10, no. 4 (October): 514–21.

Ramo, Simon. 1969b. *Cure for Chaos: Fresh Solutions to Social Problems through the Systems Approach.* New York: McKay.

Ramo, Simon. 1970. *Century of Mismatch: How Logical Man Can Reshape His Illogical Technological Society.* New York: McKay.

Ramo, Simon. 1972a. "Toward a Social-Industrial Complex." *Wall Street Journal* (February 16): 10.

Ramo, Simon. 1972b. "The Coming Shortage of Educated People." *Engineering Education* 63, no. 1 (October): 19–20.

Ramo, Simon. 1980. *America's Technology Slip.* New York: Wiley.

Ramo, Simon. 1988. *The Business of Science: Winning and Losing in the High-Tech Age.* New York: Hill and Wang.

Rasmussen, John P., ed. 1972. *The New American Revolution: The Dawning of the Technetronic Era.* New York: Wiley.

Reich, Charles A. 1970. *The Greening of America.* New York: Random House.

Reynolds, Terry S. 1983. *75 Years of Progress: A History of the American Institute of Chemical Engineers, 1908–1983.* New York: AIChE.

Reynolds, Terry S. 1991. "The Engineer in 19th-Century America." In *The Engineer in America*, Terry S. Reynolds, ed. Chicago: University of Chicago Press, 7–26.

"Research: A Policy of Protest." *Time* (February 28, 1969): 60.

Riley, Donna. 2011. "Peace Pilgrimage for a Nuclear Free World." *Reconstruct: A zine about Engineering, Social Justice, and Peace* 3: 2.

Riordan, Michael. 2006. "How Bell Labs Missed the Microchip." *IEEE Spectrum* 43, no. 12 (December): 36–41.

Roberts, Benjamin W., and Walter Lowen. 1968. "VITA: Solving Technical Design Problems throughout the World." *Engineering Education* 58, no. 7 (March): 815–17.

Robertson, James Oliver. 1980. *American Myth, American Reality.* New York: Hill and Wang.

Robin, Stanley S. 1969. "The Female in Engineering." In Perrucci and Gerstl, *The Engineer and the Social System*: 203–18.

Rodgers, Daniel T. 2011. *Age of Fracture.* Cambridge: Belknap Press of Harvard University Press.

Rojas, Fabio. 2007. *From Black Power to Black Studies: How a Radical Social Movement Became an Academic Discipline.* Baltimore: Johns Hopkins University Press.

Rosenstein, Allen B. 1961. "Engineering Science in EE Education." In *Electrical Engineering Education: Proceedings of the International Conference at Syracuse and Sagamore*, Norman Balabanian and Wilbur R. Page, eds. Syracuse: Department of Electrical Engineering, Syracuse University, 95–98.

Rosenstein, Allen B. 1968. *A Study of a Profession and Professional Education: The Final Publication and Recommendations of the UCLA Educational Development Program*. Los Angeles: School of Engineering and Applied Science, University of California.

Rosenstein, Allen B. 1970. "Professions in Modern Society." *Engineering Education* 60, no. 10 (June): 960–63.

Rosenthal, D., A. B. Rosenstein, and M. Tribus. 1961. "How Can the Objective of Engineering Education Be Best Achieved?" *Data Link* 4, no. 4 (December): 8–15.

Rostow, W. W. 1960. *The Stages of Economic Growth: A Non–Communist Manifesto*. Cambridge: Cambridge University Press.

Roszak, Theodore. 1969. *The Making of a Counter Culture: Reflections on the Technocratic Society and Its Youthful Opposition*. Garden City, NY: Doubleday.

Roy, Rustum, and Joshua Lerner. 1983. "The Status of STS Activities at U.S. Universities." *Bulletin of Science, Technology and Society* 3: 417–32.

Rybczynski, Witold. 1980. *Paper Heroes: A Review of Appropriate Technology*. Garden City, NY: Anchor Books.

Sakellariou, Nicholas. "Engineers and Sustainability: Between Invisibility and (Yet Another) Story of Contextualization in a Period of Crisis." Unpublished manuscript.

Schatzberg, Eric. 2006. "*Technik* Comes to America: Changing Meanings of *Technology* before 1930." *Technology and Culture* 47, no. 3 (July): 486–512.

Schevitz, Jeffrey M. 1979. *The Weaponmakers: Personal and Professional Crisis during the Vietnam War*. Cambridge: Schenkman.

Schlesinger, Arthur M. Jr. 1970. "The Velocity of History." *Newsweek* (July 6): 32–34.

Schön, Donald A. 1967. *Technology and Change: The New Heraclitus*. New York: Delacorte.

Schön, Donald A. 1969. "The Diffusion of Innovation." *Innovation* 6 (October): 42–53.

Schön, Donald A. 1971. *Beyond the Stable State*. New York: Random House.

Schulman, Bruce J. 1991. *From Cotton Belt to Sunbelt: Federal Policy, Economic Development, and the Transformation of the South, 1938–1980*. New York: Oxford University Press.

Schumacher, E. F. 1973. *Small Is Beautiful: Economics as If People Mattered*. New York: Harper and Row.

Schwartz, Rebecca Press. 2000. "Re-evaluating Sponsored Research at Princeton: The Kuhn Committee and IDA." Unpublished manuscript, last modified April 10.

Schweber, S. S. 2000. *In the Shadow of the Bomb: Bethe, Oppenheimer, and the Moral Responsibility of the Scientist*. Princeton: Princeton University Press.

Seely, Bruce E. 1993. "Research, Engineering, and Science in Engineering Colleges, 1900–1960." *Technology and Culture* 34, no. 2 (April): 344–86.

Seely, Bruce E. 1995. "SHOT, the History of Technology, and Engineering Education." *Technology and Culture* 36, no. 4 (October): 739–72.

Seely, Bruce E. 1999. "The Other Re-engineering of Engineering Education, 1900–1965." *Journal of Engineering Education* 88, no. 3 (July): 285–94.

Segal, Howard P. 1994. *Future Imperfect: The Mixed Blessings of Technology in America*. Amherst: University of Massachusetts Press.

Seron, Carroll, and Susan Sibley. 2009. "The Dialectic between Expert Knowledge and Professional Discretion: Accreditation, Social Control, and the Limits of Instrumental Logic." *Engineering Studies* 1, no. 2 (July): 101–28.

Sinclair, Bruce. 2001. "Local History and National Culture: Notions on Engineering Professionalism in America." In *American Technology*, Carroll W. Pursell Jr., ed. Malden, MA: Blackwell, 145–54.

Singularity University. 2011. *Exponential Technologies Executive Program*. Moffett Field, CA: Singularity University.

Slaby, Steve M. 1952. *The Labor Court in Norway*. Oslo: Norwegian Academic Press.

Slaby, Steve M. 1970. "Technology and Society: Present and Future Challenges." In *Man-Society-Technology: Representative Addresses and Proceedings of the American Industrial Arts Association 32 Annual Convention*, Linda A. Taxis, ed. Washington, DC: American Industrial Arts Association, 9–11.

Slaby, Steve M. 1979. "What Should We Ask of the History of Technology?" In *The History and Philosophy of Technology*, ed. George Bugliarello and Dean B. Doner, eds. Urbana: University of Illinois, 112–27.

Slaton, Amy E. 2010. *Race, Rigor, and Selectivity in U.S. Engineering: The History of an Occupational Color Line*. Cambridge: Harvard University Press.

Slayton, Rebecca. 2003. "Speaking as Scientists: Computer Professionals in the Star Wars Debate." *History and Technology* 19, no. 4: 335–64.

Smiles, Samuel. 1861. *Lives of the Engineers, with an Account of Their Principal Works: Comprising Also a History of Inland Communication in Britain*, vol. 1. London: Murray.

Smith, Alice Kimball. 1970. *A Peril and a Hope: The Scientists' Movement in America, 1945–47*. Cambridge: MIT Press.

Smith, Merritt Roe, and Leo Marx, eds. 1994. *Does Technology Drive History? The Dilemma of Technological Determinism*. Cambridge: MIT Press.

Smith, Ralph J. 1969. *Engineering as a Career*. New York: McGraw-Hill.

Smits, Alexander J., and Courtland D. Perkins. 2004. "Aerospace Education and Research at Princeton University, 1942–1975," Accessed October 1, 2011. http://www.princeton.edu/mae/aboutus/history-1/.

Snyder, Benson R. 1970. *The Hidden Curriculum*. New York: Knopf.

Starr, Chauncey. 1969. "Social Benefit versus Technological Risk." *Science* 165, no. 3899 (September 19): 1232–38.

Staudenmaier, John M. 1985. *Technology's Storytellers: Reweaving the Human Fabric*. Cambridge: SHOT and MIT Press.

Stephan, Karl D. 2006. "Notes for a History of the IEEE Society on Social Implications of Technology." *IEEE Technology and Society Magazine* 25, no. 4: 5–14.

Stevens, William K. 1969. "Study Terms Technology a Boon to Individualism." *New York Times* (January 18): 1, 17.

Stoff, Joshua. 1993. *Picture History of World War II American Aircraft Production*. New York: Dover.

"A Student of Technology: Emmanuel George Mesthene." *New York Times* (January 18, 1969): 17.

Students for a Democratic Society. 1962. *The Port Huron Statement*. New York: SDS.

Subcommittee on Science, Research and Development of the Committee on Science and Astronautics. 1969. *Technological Assessment—1969*. Washington, DC: US Government Printing Office.

"Supersonic Counterattack." *Time* (March 22, 1971): 15.

Susskind, Charles. 1973. *Understanding Technology*. Baltimore: Johns Hopkins University Press.

Taviss, Irene. 1972. A Survey of Popular Attitudes Toward Technology. *Technology and Culture* 13, no. 4 (October): 606–21.

Taviss, Irene, and Linda Silverman. 1969. *Technology and Values*. Cambridge: Harvard University Program on Technology and Society.

Tichi, Cecelia. 1987. *Shifting Gears: Technology, Literature, Culture in Modernist America*. Chapel Hill: University of North Carolina Press.

"Top Idea Men Trade Ideas." *Business Week* (January 31, 1970): 32–33.

Turner, Fred. 2006. *From Counterculture to Cyberculture: Stewart Brand, the Whole Earth Network, and the Rise of Digital Utopianism*. Chicago: University of Chicago Press.

Turner, Fred. 2008. "Romantic Automatism: Art, Technology, and Collaborative Labor in Cold War America." *Journal of Visual Culture* 7, no. 1: 5–26.

US Bureau of Labor Statistics. 1963. *Scientists, Engineers, and Technicians in the 1960's: Requirements and Supply*. Washington, DC: US Government Printing Office.

Unger, Stephen H. 1971. "New Engineering Conference: CSRE holds Counterconference during IEEE Convention." *Spark* 1, no. 2 (fall): 2–5.

Unger, Stephen H. 1973a. "The BART Case: Ethics and the Employed Engineer." *IEEE CSIT Newsletter* 4 (September): 6–8.

Unger, Stephen H. 1973b. "Codes of Engineering Ethics." *IEEE CSIT Newsletter* 5 (December): 1, 4–5.

Unger, Stephen H. 1994. *Controlling Technology Ethics and the Responsible Engineer*, 2nd ed. New York: Wiley.

Veblen, Thorstein. 1921. *The Engineers and the Price System*. New York: Huebsch.

"Village Technology." *Whole Earth Catalog* (fall 1968): 18.

Volunteers in Technical Assistance. 1975. *A Guide to Mobilizing Technical Assistance Volunteers: A "How-to" Manual for Organizing Local Volunteer Assistance for Community Problem Solving*. Mt. Rainier, MD: VITA.

Von Fange, Eugene K. 1959. *Professional Creativity*. Englewood Cliffs, NJ: Prentice-Hall.

Vonnegut, Kurt, Jr. 1952. *Player Piano*. New York: Delacorte Press.

Wachs, Theodore Jr. 1964. *Careers in Engineering*. New York: Walack.

Waldman, Theodore. 1971. "Quest for Commonwealth, an Applied Humanities Course." In *Applying Humanities to Engineering*. Claremont, CA: Harvey Mudd College, 1–11.

Walker, Charles R. 1962. *Modern Technology and Civilization: An Introduction to Human Problems in the Machine Age*. New York: McGraw-Hill.

Walker, Eric A. 1972. "Engineers: A Time for Leadership." In *Engineering, Technology and Society: The Gwilym A. Price Engineering Lectures*, ed. H. E. Hoelscher. Pittsburgh: University of Pittsburgh, 37–48.

Wang, Jessica. 1999. *American Science in an Age of Anxiety: Scientists, Anticommunism, and the Cold War*. Chapel Hill: University of North Carolina Press.

Wang, Zuoyue. 2008. *In Sputnik's Shadow: The President's Science Advisory Committee and Cold War America*. New Brunswick, NJ: Rutgers University Press.

Ward, Barbara. 1966. *Spaceship Earth*. New York: Columbia University Press.

Ward, Barbara. 1962. "We Are All Developing Nations." *New York Times Magazine* (February 25): 4, 38, 40, 43.

Weinstein, Alvin S., and Stanley W. Angrist. 1970. *An Introduction to the Art of Engineering*. Boston: Allyn and Bacon.

Wiener, Norbert. 1954. *The Human Use of Human Beings: Cybernetics and Society*. Boston: Houghton Mifflin.

Wiesner, Jerome B. 2003. "Letter to Howard Johnson, April 3, 1967." In *Jerry Wiesner: Scientist, Statesman, Humanist: Memories and Memoirs*, Walter A. Rosenblith. Cambridge: MIT Press, 473–75.

Westinghouse Electric Company. 1967. "Go Westinghouse, Young Man!" *Engineering and Science* 30, no. 4 (January): 1.

Whalley, Peter. 1986. *The Social Production of Technical Work: The Case of British Engineers*. Albany: State University of New York Press.

Whyte, William H. Jr. 1956. *The Organization Man*. New York: Simon and Schuster.

Williams, Rosalind. 2002. *Retooling: A Historian Confronts Technological Change*. Cambridge: MIT Press.

Williamson, Bess. 2008. "Small Scale Technology for the Developing World: Volunteers for International Technical Assistance, 1959–1971." *Comparative Technology Transfer and Society* 6, no. 3 (December): 236–58.

Winner, Langdon. 1977. *Autonomous Technology: Technics-out-of-Control as a Theme in Political Thought*. Cambridge: MIT Press.

Wisely, William H. 1974. *The American Civil Engineer, 1852–1974: The History, Traditions, and Development of the American Society of Civil Engineers, Founded 1852*. New York: ASCE.

Wisnioski, Matthew. 2003. "Inside the System: Engineers, Scientists, and the Boundaries of Social Protest in the Long 1960s." *History and Technology* 19, no. 4: 313–33.

Wisnioski, Matthew. 2005. "Engineers and the Intellectual Crisis of Technology." PhD dissertation. Princeton University.

Wisnioski, Matthew. 2009. "'Liberal Education Has Failed': Reading Like an Engineer in 1960s America." *Technology and Culture* 50, no. 4 (October): 753–82.

Wittner, Lawrence S. 1997. *Resisting the Bomb: A History of the World Nuclear Disarmament Movement, 1945–1970*. Stanford: Stanford University Press.

Wolfle, Dael. 1967. "The Big Story—The Meaning of Science and Technology in Modern Society. Do the Engineering Society Publications Cover It?" In *Engineering Societies and Their Literature Programs*. Larry Resen, ed. New York: EJC, 63–66.

Zimmerman, Bill, Len Radinsky, Mel Rothenberg, and Bart Meyers. 1972. *Towards a Science for the People*. Boston: New England Free Press.

Zussman, Robert. 1985. *Mechanics of the Middle Class: Work and Politics among American Engineers*. Berkeley: University of California Press.

Name Index

Johns, Jasper, 142
Johnson, Dan, 128–31
Johnson, Howard, 101, 180
Johnson, Lyndon, 80, 154, 181
Jünger, Friedrich Georg, 76

Kafka, Franz, 159
Kaiser, David, 28
Kantrowitz, Arthur, 38
Kaplan, Irving, 182
Kaysen, Carl, 52
Keck, James, 103
Keil, Alfred A. H., 181
Kemp, Jack, 120
Kemper, John Dustin, 28–29, 32, 187
Kennedy, John F., 151, 181
Kerr, Clark, 169
Kheel, Theodore W., 143
King, Martin Luther, Jr., 43
Kinzel, Augustus, 60
Klein, Burton H., 174
Klüver, Billy, 138, 140, 142–44, 148, 230n71
Knowlton, Ken, 145
Koch, Ed, 1
Koestler, Arthur, 42
Kohl, Walter H., 61–62
Koning, Hendrik B., 82
Kranzberg, Melvin, 42, 60, 86, 168
Kuhn, Thomas, 110–11
Kurzweil, Ray, 196

Landau, Ralph, 9–10
Lavoisier, Antoine-Laurent, 6, 14
Layton, Edwin T., Jr., 4–5, 7, 19–21, 70, 93
 engineers' reading of, 5, 86, 200n18
Leach, H. W., 161
Leary, Timothy, 36
Leavitt, William, 124
Leslie, Stewart W., 23–24
Lettvin, Jerome, 36
Leventman, Paula Goldman, 121
Lewis, Warren K., 178, 181
Light, Jennifer S., 8
Little, W. E., 72
Locke, John, 164

Luria, Salvador, 181
Lyttle, Brad, 112, 225n68

Maass, William G., 151, 154–55
Macdonald, Dwight, 71
Malcolm X, 209n76
Marcuse, Herbert, 3, 43, 45–46, 98
 engineers' reading of, 42, 47–49, 100, 110, 114, 117
 and Harvard University Program on Technology and Society, 63
Marine, Gene, 18, 42, 168
Mark, Robert, 167
Marlowe, Donald E., 5, 70, 76–78, 82, 86–87, 89–91
Marx, Leo, 9, 12
McAllister, Don, 37
McCracken, Paul, 120
McDermott, John, 42, 63, 148, 17
McGovern, George, 121
McLuhan, Marshall, 51, 126
 engineers' reading of, 42, 143, 153
 and humane technology, 158–59
McMahon, Darrin M., 14
McPherson, J. H., 127
Mee, Thomas R., 146
Melman, Seymour, 1, 97–98, 116, 152
Menocchio, 1
Merton, Robert K., 53, 215n90
Mesthene, Emmanuel G., 52, 153, 157, 190, 212n44
 and engineering education, 164, 166–67, 177
 and Harvard University Program on Technology and Society, 52–54, 63– 64, 157, 213n53
 and Innovation Group, 149, 155
 and Society for the History of Technology, 60
 engineers' reading of, 61, 144, 231n80
Metcalfe, Bob, 196–97
Michelangelo, 15
Mills, C. Wright, 28, 68
Moore, John R., 61

Moore, Kelly, 7
Morgan, Robert P., 133
Morison, Elting E., 182
Morton, Jack A., 151, 155–57,
  233n122
Mumford, Lewis, 3–4, 45–46, 98,
  128
  and engineering education, 80, 158,
  159, 164–65, 181, 240n69
  and Harvard University Program on
  Technology and Society, 63–64
  engineers' reading of, 6–7, 41–42,
  47, 60, 76, 100
Muñoz, David, 194
Murray, Chris, 122
Muste, A. J., 79

Nader, Ralph, 43, 219n60
Nakaya, Fujiko, 146
Nason, Howard K., 62
Nelson, Wilbur, 120
Neswald, Ron, 156
Nevill, Gale E., Jr., 166
Niebuhr, H. Richard, 71
Nikolsky, Alexander, 107
Nixon, Richard, 2, 110, 134, 136,
  158, 169, 174
Noble, Daniel E., 27–28
Noble, David F., 19–21, 93, 191

O'Connor, John A., 166
Ogburn, William Fielding, 9, 50, 60,
  211n33, 214n66
Oglesby, Carl, 178, 181
Oldenziel, Ruth, 9, 20–21
Olmsted, Sterling P., 161, 164–65
Orwell, George, 42

Packard, David, 1–4, 11, 74
Palmer, James D., 135
Paschkis, Victor, 1, 11, 72, 78–82,
  191
  and American Society of Mechanical
  Engineers, 72, 76–77, 82–84,
  89–91, 96
  and Committee for Social
  Responsibility in Engineering, 1

and Committee on Social
  Implications of Technology, 117
Pauling, Linus, 79
Perkins, Courtland, 107
Perrucci, Robert, 207n48, 208n63
Petroski, Henry, 235n144
Picht, Georg, 76
Piel, Gerard, 53
Pierce, John R., 146
Pilisuk, Marc, 98
Plato, 164
Platt, Joseph B., 176, 185
Price, Don K., 51, 52, 211n36
Probstein, Ronald F., 103–106
Proteus, Paul, 123, 159
Proxmire, William, 37
Pursell, Carroll W., Jr., 168

Rae, John, 168, 176
Ramo, Simon, 11–12, 55–59, 60, 189,
  190, 214n66
  and engineering education, 58, 166,
  177
  and J. Herbert Hollomon, 153
  and "social-industrial complex," 58,
  119
Rand, Ayn, 42
Rauschenberg, Robert, 142–43
Reagan, Ronald, 57
Reich, Charles, 42, 44, 47
Reuther, Walter P., 53
Rhodes, Allen F., 62
Rhodes, Joseph, Jr., 161, 174, 185
Ricardo, David, 86
Rickover, Hyman G., 124
Riley, Donna, 192–93
Roe, Kenneth A., 87
Rosenstein, Allen B., 171–73, 239n50
Rosenthal, Daniel, 171
Roszak, Theodore, 4, 42, 89, 180
Rulseh, Roy, 64
Ryker, John, 72

Salvadori, Mario, 89
Sayre, Daniel, 107
Schlafly, Phyllis, 209n76
Schlesinger, Arthur M., Jr., 41

Subject Index

Acceleration Studies Foundation, 196

Aerospace industry, 22–23, 36, 55–56, 105, 152. *See also* Defense, National; Labor, Engineering
conversion of, 96, 103–106, 118–20, 153
response to technology's critics by, 34, 61, 93, 105, 119–20, 124

American Institute of Aeronautics and Astronautics (AIAA), 60, 103, 119

American Institute of Chemical Engineers (AIChE), 10, 90, 132

American Society of Civil Engineers (ASCE), 6, 20, 64–65, 89–91, 159, 191, 194

American Society for Engineering Education (ASEE), 70, 73, 161–65, 179, 236n12

American Society of Mechanical Engineers (ASME), 5–6, 62, 70–78, 82–93, 194, 203n16
ASME Goals, 84–90, 219n55, 219n58, 219n60
Technology & Society Committee, 6, 67, 82–84, 96, 117

American Telephone and Telegraph Company (AT&T), 20, 24, 49, 155

Appropriate technology (AT) movement, 124–38, 188, 194. *See also* Engineers Without Borders; Humanitarian engineering; Volunteers for International Technical Assistance

AT International, 137

Cooperative for American Relief Everywhere, 131

Intermediate Technology Development Group, 137

Maryknoll, 131

National Center for Appropriate Technology, 137

Private Agencies Cooperating Together, 137

Aquarius Project, 115

Association of Technical Professionals, 98

Automation, 43, 50–51, 80, 123
and American Foundation on Automation and Unemployment, 143
and Automation House, 143, 146, 148
and *Triple Revolution*, 80, 218n39

Autonomous technology, 3–4, 11–14, 38, 41–65, 125–28, 148–51. *See also* Technological change, ideology of; Technological politics, ideology of
and engineering education, 162–74, 184–85
and engineering professionalism, 68–69, 76, 80–84, 91–93, 153–54
and military-industrial complex, 96–101

Service (cont.)
  disillusionment with, 33–39, 102,
    109–11, 114, 122
  to government, 27, 31, 86, 178
  to multiple parties, 20–21, 68,
    95–96
  and professionalism, 19–21, 31–32,
    68–71
  vs. servitude, 96–98, 111
Singularity University, 196–97
Sloan Foundation, 132, 177, 183
Society for the History of Technology
  (SHOT), 60
Society for Social Responsibility in
  Science (SSRS), 79
Society of Women Engineers, 34
"Socio-technical mismatch," 11–12,
  56–59, 189
Socio-technologist, 54–59, 111,
  161–85
Soviet Union, 29, 38, 43, 50, 95, 152
Space program. *See also* Aerospace
  industry; National Aeronautics and
  Space Administration
  and character of engineering, 34, 74,
    103, 105, 120–21, 169
  and critiques of technology, 33, 38,
    63
  and engineering recruitment, 29, 31
  and systems engineering, 24, 56
Spark (CSRE), 1–2, 96–97, 112–17,
  122
Sponsored research. *See under*
  Defense, national
Stanford University, 15, 24, 184
Students for a Democratic Society
  (SDS), 14, 41, 98, 102, 110, 178
Supersonic transport (SST), 14, 37,
  105, 120, 174
Sustainability, 191, 194, 196
Syracuse Peace Council, 122
Systems approach, 24, 56–58, 191
  and out-of-control technology, 52,
    56–58, 61, 86–87, 165, 181–82,
    188
  and interdisciplinarity, 128, 165–67,
    171–72, 176, 181

Technological change, ideology of,
  11–13, 44, 49–65, 189–90, 196–97.
  See also Autonomous technology
  and adaptation/adjustment,
    professional, 62, 91, 144, 190, 196
  and adaptation/adjustment, societal,
    11–12, 44, 50–53, 127, 169
  and culture lag, 50, 60, 64, 197,
    211n33 (*see also* "Socio-Technical
    Mismatch")
  and engineering education, 58–59,
    166–69, 174–75, 177, 184–85
  and expert anticipation/control,
    11–12, 44, 51–54, 58–61, 83, 169,
    181, 197
  and governance, 50, 54, 58–59, 61,
    153–54, 189–90
  and humane technology, 52, 63,
    128, 143, 153–58
  and an ideology of technological
    politics, 45, 49, 51, 59–60, 63–65,
    196
  and inevitability, 12, 50, 52–53, 58,
    91, 96, 121, 144, 158, 161, 190,
    196
  and keeping pace, 50, 93, 125, 157,
    189–90, 197
  as a middle philosophy, 52, 65, 76
  and professionalism, 58–59, 62, 69,
    76, 80–93
  and responsibility, 51, 60, 62–63,
    76, 80–93, 117
  and unintended effects, 11, 44, 50,
    58, 69–70, 76, 80–83, 88, 90–91,
    117, 143, 166–67
Technological determinism, 12,
  52–53, 128, 201n37
Technological politics, ideology of,
  44–49, 63–65, 98, 188–89, 196.
  See also Autonomous technology;
  Critical theories of technology;
  Humane technology
  and appropriate technology, 125,
    135
  and design, 47–49, 144
  and engineering dissent, 49, 76–77,
    109–10, 114–15, 120–22, 192–93